Contract Practice for Surveyors

WITHDRAWN

D0233680

Contract Practice for Surveyors

FOURTH EDITION

Jack Ramus, MRICS

Simon Birchall,

MBA, BSc(Hons), MRICS, PGCE

AND

Phil Griffiths

LONDON AND NEW YORK

First published by Butterworth-Heinemann

First edition 2006

This edition published 2011 by Spon Press
2 Park Square, Milton Park, Abingdon, Oxon OX14 4RN

Simultaneously published in USA and Canada
by Taylor & Francis Group, 711 Third Avenue, New York, NY 10017, USA

Spon Press is an imprint of the Taylor & Francis Group, an informa business

Copyright © 2006, Phil Griffiths, Simon Birchall and Jack Ramus.
All rights reserved

The right of Phil Griffiths, Simon Birchall and Jack Ramus to be identified
as the authors of this work has been asserted in accordance with the Copyright,
Designs and Patents Act 1988

No part of this publication may be reproduced, stored in a retrieval system
or transmitted in any form or by any means electronic, mechanical, photocopying,
recording or otherwise without the prior written permission of the publisher

Notice
No responsibility is assumed by the publisher for any injury and/or damage to persons
or property as a matter of products liability, negligence or otherwise, or from any use
or operation of any methods, products, instructions or ideas contained in the material
herein. Because of rapid advances in the medical sciences, in particular, independent
verification of diagnoses and drug dosages should be made

British Library Cataloguing in Publication Data
A catalogue record for this book is available from the British Library

Library of Congress Control Number: 2006927668

ISBN: 978-0-7506-6833-0

Transferred to Digital Printing in 2013

Contents

Figures and tables

Figures

Tables

Preface

Since publication of the third edition in 1996 there have been many developments in construction procurement, finance, and contract administration. During the latter part of the twentieth century developments in Information Technology and electronic communications have influenced many of the traditional procedures within the construction industry. Since 1996 there have been wide ranging changes in many of the standard forms of contract used in the construction industry as demonstrated by the JCT who have undertaken two major revisions of their contract documentation initially in 1998 and again in 2005. These changes have been incorporated in this fourth edition to ensure the book reflects current industry practice. The book has also been brought up to date with current legislation, for example the Housing Grants, Construction and Regeneration Act and the Enterprise Act.

During the preparation of the previous editions the scope of the book had been gradually widened to encompass new and relevant areas of contract administration. Value Management has been added to this edition to complement the material from previous publications. As a result of this broader approach to contract administration we feel that this book will be of use to quantity surveyors, building surveyors and general practice surveyors and will also be useful to any professional engaged in the field of property or construction.

The book provides a useful reference for degree students, graduates, practitioners engaged in all aspects of construction contracts, and candidates sitting membership examinations of the various construction industry professional associations. We hope that the detailed examples will continue to meet the need for clear guidance in the practical procedures relating to contract practice.

The first two editions of the book were written principally for those employed in private practice or in central or local government departments. However, it is recognised that contractors' surveyors are now playing a more important role in the area of contract administration as reflected in the current JCT contracts. The fourth edition reflects this changing role and provides a balanced view of the work undertaken by both the client's and contractor's surveyor. Throughout the book, where specific reference is made to surveyors in construction companies, the term 'contractor's surveyor' is used; where reference is to surveyors in general, the term 'the surveyor' is used.

Attention is drawn to the fact that the text relates only to English law and no reference is made to differences in law and practice applicable in Scotland. It is hoped nonetheless that Scottish surveyors will find the book of use and value in their studies and/or employment.

We wish to acknowledge the help and encouragement we have received in the preparation of this fourth edition. We thank Sweet & Maxwell Ltd for permission to reproduce extracts from the JCT Standard Building Contract With Quantities; to the Office of Public Sector Information for their assistance regarding the use of Crown copyright material: Price Adjustment Formulae for Construction Contracts, Users Guide, 1990 Series of Indices; Price Adjustment Formulae for Construction Contracts Monthly Bulletin of Indices; Department of Health Expenditure Forecasting Formula. With reference to these documents we acknowledge Crown copyright material is reproduced with the permission of the Controller of HMSO and the Queen's Printer for Scotland.

Jack Ramus
Simon Birchall
Phil Griffiths
May 2006

1

Contracts and the surveyor

The nature and form of contracts

All aspects of the surveyor's professional work relate directly or indirectly to the costs of construction work of all kinds. This includes new building and civil engineering projects, and the alteration, extension or refurbishment of existing buildings. In addition, his expertise extends to consequential costs incurred by owners and occupiers after the completion of construction projects. These include repairs and redecoration, replacement of components and the operating costs of installations such as lifts, air-conditioning, heating, etc. All these costs are known collectively as 'costs-in-use' or 'life cycle costs'.

Such costs are incurred when building and engineering contractors of all kinds carry out work for clients of the construction industry. This necessarily requires that the parties concerned in a particular project enter into a contract, i.e., a legal agreement to discharge certain obligations. So, as a simple example, a builder enters into a contract with the owner of a plot of land in which he undertakes to build a house complying with the owner's requirements. On his part, the owner undertakes to pay the builder a stated sum of money.

In its simplest form a contract may be an oral agreement, the subsequent actions of the parties providing evidence of the existence of a contract. Such an agreement is legally binding, provided it does not entail an illegal act.

Alternatively, a contract may be set out in a letter or other brief document stating the essential matters agreed upon. However, such forms of contract are only suitable for the simplest projects, such as building a garage onto an existing house or erecting fencing around a plot of land. Even in those cases disputes may arise, leading to arbitration or litigation, both costly processes.

In the case of most building and engineering projects, therefore it is essential that the conditions under which a contract is to be carried out and the obligations, rights and liabilities of the parties should be fully and clearly stated in writing. Even then, disputes can still arise because of ambiguity or omission in the contract documents or because something unforeseen arises during the period of the contract.

For most projects, printed standard forms are used as the basis of the contract. These set out the conditions applicable to the contract and cover all the matters which experience has shown need to be specified in order that both parties may have a clear picture of their rights and obligations.

Those matters, which are peculiar to a particular project such as names of the parties, the contract sum, commencing and completion dates, etc. are incorporated into the contract by insertion into blank spaces in the Articles of Agreement or the Contract Particulars, both of which are part of the printed document. It is sometimes necessary for the printed text of a clause of a standard form to be amended, added to or deleted in order to accommodate particular requirements of a building owner or because the standard wording does not suit the particular circumstances of a project. Such amendments must be made with care, to avoid introducing ambiguities or inconsistencies with other clauses or other contract documents.

The National Joint Consultative Committee for Building (NJCC) issued a Procedure Note[1] advising against extensive amendments to the Joint Contracts Tribunal (JCT) Form, on the grounds that such amendments may introduce inequity and tend away from greater standardization of building procedures. The Note also advises that when amendments are considered necessary, only a competent person should amend the Form so that ambiguities or inconsistencies, which might lead to difficulties later, are not introduced. This advice applies equally to all standard forms. This advice is reiterated by the JCT where they warn that 'ill-conceived amendments can produce unintended results'.[2] There have been instances where the client's advisers have altered a standard form of contract to the benefit of the client only to find their amendment to be worthless because they had deleted a clause in isolation without giving proper consideration to other relevant clauses that remained in the contract *(Wates Construction Ltd v Franthom Property Ltd, 1991, 53 BLR 23).*

The advantages of using standard forms of contract instead of composing a contract document afresh for each project are:

(a) because they are written by legal experts, ambiguities and inconsistencies are reduced to a minimum;
(b) the rights and obligations of both parties are set out clearly and to the required degree of detail;
(c) because of standardization, frequent users are familiar with the provisions in a particular Form, resulting in a greater degree of consistency in their application;
(d) the time and expense which would otherwise be incurred in preparing a fresh document for each occasion is avoided; and
(e) over a period of time a history of case law is built up as ambiguities and disputes are settled through litigation, which can provide a good source of knowledge for construction professionals.

The standard forms, most widely used for building projects, are those in the range produced by the JCT, a body which is composed of representatives of the following organizations:

- Association of Consulting Engineers
- British Property Federation
- Construction Confederation
- Local Government Association
- National Specialist Contractors Council
- Royal Institute of British Architects
- Royal Institution of Chartered Surveyors
- Scottish Building Contracts Committee.

Some of the JCT 2005 edition standard forms of contract and sub-contract are listed below and dealt with in more detail in Chapter 3:

1 Standard Building Contract
   (a) With Quantities (SBC/Q).
   (b) With Approximate Quantities (SBC/AQ).
   (c) Without Quantities (SBC/XQ).
2 Design and Build Contract (DB).
3 Intermediate Building Contract (IC).
4 Management Building Contract (MC).

5 Management Works Contract Conditions (MCWK/C).
6 Construction Management Agreement (CM/A).
7 Construction Management Trade Contract (CM/TC).
8 Major Project Construction Contract (MP).
9 Prime Cost Building Contract (PCC).
10 Minor Works Building Contract (MW).
11 Measured Term Contract (MTC).
12 Named Sub-Contract Conditions to Intermediate Contract (ICSub/NAM/C).
13 Standard Building Sub-Contract Conditions to the Standard Building Contract (SBCSub/C).

In addition to the above, the JCT produces a number of alternative contracts incorporating contractor's design, as well as various ancillary documents and Practice Notes relating to JCT Forms. A complete catalogue of JCT publications is available from the publishers Sweet & Maxwell Ltd, or from the JCT website.

Other standard forms of contract are:

(a) ICE Conditions of Contract (for civil engineering works, published by the Institution of Civil Engineers and the Federation of Civil Engineering Contractors).
(b) Engineering and Construction Contract produced by the ICE.
(c) General Conditions of Government Contracts for Building and Civil Engineering Works. The most common of these forms is known as GC/Works/1, but there are at least another dozen Government standard contracts designed for a variety of works and procurement routes.

The above list is not exhaustive. See Chapter 3 for further information about non-JCT Forms of Contract.

The surveyor's role

Many of the provisions in standard forms of contract relate to financial matters, and surveyors need to have a detailed knowledge and understanding of them and be able to apply those provisions properly and in a professional manner. Consequently, most surveyors have acquired considerable expertise in the use of standard forms of contract and many architects look to the surveyors with whom they regularly work for advice and guidance on contractual matters. Frequently, it falls to the surveyor to prepare the contract documents ready for the execution of the contract by the parties, a duty which requires particular care and attention to detail.

In the performance of his duties during and after the construction period of the project, the surveyor has the duty to ensure that all actions taken in relation to the financial administration of the contract will be fair to both the parties. While there is often, and should be, such an attitude underlying the surveyor's approach to his duties, he must not forget two important facts. First, 'fairness' is often a subjective matter so that an action or decision may be seen as fair by one person and as very unfair by another. Second, the surveyor's authority to act rests ultimately upon the terms of the contract and, consequently, his actions must always be within the scope of those terms. Both parties to a contract, Employer and contractor, are bound by the terms of that contract, having been freely agreed to them when they signed the contract. Should any one of those terms in its practical application become disadvantageous, or even financially disastrous, to either party, the surveyor has no authority to vary its provisions in order to alleviate those consequences.

The young surveyor may be tempted, by a sense of fair play which in other circumstances would be very commendable, to act in a particular situation in a way which seems to him to be fairer to one or other of the parties rather than in accordance with the strict application of the contract terms. For example, it might seem very unfair that a contractor, having complied with a verbal instruction by the architect to carry out a variation to the design of a building, should be refused payment for the extra cost just because the variation had not been confirmed in writing. The surveyor cannot do otherwise, however unfair it may seem, if the contract requires that all variations, to be valid, must be in writing.[3] Similarly, if the surveyor happens to know during the construction period that either party is having cash flow problems, that knowledge must not be allowed to influence the value he puts upon work completed or unfixed materials on site when preparing a valuation for an interim certificate. However sympathetic he may feel towards either party, the surveyor must not allow his personal feelings to prevail over his judgement as to the proper application of any term of a contract. He must remember that any unwarranted advantage given to one party is likely to be at the same time an equal disadvantage to the other. The foregoing remarks apply particularly to surveyors employed in private professional firms and in the offices of local authorities, government departments, and other public bodies.

The surveyor employed by a building contractor is in a somewhat different situation in that his responsibility and loyalty is primarily to his employer. Nevertheless, his actions on behalf of the contractor must equally be governed by the terms of each contract with which he deals. There is no reason, however, why the contractor's surveyor should not invoke a term of contract to the advantage of his employer and there can be no objection to his doing so. For example, by a careful comparison of the bills of quantities and the drawings, he may be able to detect errors of omission in the quantities, the correction of which may be provided for in the contract.[4] On the other hand, he must recognize that a claim, of whatever kind, can only be justified if it is based upon a specific term of the contract and, even if shown to be valid, the valuation of it may only be reached by prescribed means.[5]

But in whatever kind of organization a surveyor works, his attitude to carry out his duties should be a completely professional one. That is to say, the quality of the work he does should be as high as he is able to achieve, without becoming pedantic. Despite the pressures of a highly competitive business world, as a member of a respected and worthy profession, the surveyor should aim to be entirely honest and impartial in his interpretation and application of the terms of a building contract, on whichever side of the contractual 'fence' he may be.

The surveyor's post-contract duties

The following is a summary of the duties normally carried out by the surveyor following acceptance of a tender for a project where an architect is responsible for the supervision and management of the contract. Some of course may not be applicable to particular projects. The respective chapters in which those duties are described in detail are given in brackets.

Before work starts on site

1 Arrange for contract documents to be prepared ready for signatures of the parties (Chapter 6).
2 Prepare forecast of 'rate of spend' during the construction period and advise client on anticipated liability for payments

on account to the contractor, giving dates and amounts (Chapter 12).

3 Make preliminary arrangements for preparing valuations for payments on account in consultation with the contractor's surveyor; analyse preliminaries and calculate amounts of time-related payments and percentage rate of cost-related payments; prepare schedule for stage payments, if applicable (Chapter 10).

During construction period

4 Prepare valuations for payments on account at the intervals stated in the contract and agree with contractor's surveyor (Chapter 10).
5 Plot payments on account on 'rate of spend' graph and report to architect on any significant divergences (Chapter 12).
6 Prepare estimates of likely cost of variations on receipt of copies of architect's instructions; later measure and value; check and price daywork vouchers (Chapter 7).
7 Advise architect, if requested, on expenditure of provisional sums; measure and value work carried out by the main contractor against provisional sums (except where lump sum quotations have been accepted) and adjust contract sum accordingly (Chapter 11).
8 Prepare financial reports for architect and client at the time of interim payments (Chapter 12).
9 Check the main contractor's notifications for changes in levies, contributions and taxes, etc., if applicable; alternatively, apply price adjustment indices to amounts included in interim valuations (Chapters 8).
10 Measure projects based on schedules of rates or on bills of approximate quantities as the work proceeds, either on site or from architect's drawings, and value at contract rates.
11 Advise architect, if requested, on contractor's claims (if any) for loss and expense payments; if accepted, negotiate claims with contractor (Chapter 9).

After construction period

12 Prepare final account and agree details and total with contractor's surveyor (Chapter 11).

Duties of a contractor's surveyor

The duties of a contractor's surveyor can be very varied, a surveyor working for a small company may be responsible for all aspects of a project from inception to completion whereas larger construction firms tend to have specialist departments for various aspects of contract administration, in which case a surveyor may find himself working on a specific or limited area of project administration. The role of a contractor's surveyor can be quite pressurized as they are frequently responsible for the financial performance of a project and have to deal with the client's advisers to try and ensure the construction company gets the best return from the project. At the same time the surveyor is likely to have extensive dealings with sub-contract organizations who will be carrying out the bulk of the work and will have to manage their demands for payment, claims and extras. In this role the surveyor will be trying to maximize his company's inflow of cash while trying to minimize the outflow, thereby protecting the contractor's cash flow and hopefully ensuring that the anticipated profit margins are met. The tasks that may be carried out by a contractor's surveyor are as follows:

1. Review tender documentation, especially checking non-standard forms of contract.
2. Prepare estimates.
3. Prepare ordering schedules.
4. Place orders for materials and sub-contracts.
5. Prepare and organize bonus schemes.
6. Check bills of quantities for errors and carry out remeasures.
7. Monitor cost control systems (Chapter 12).
8. Prepare cost/value reconciliations (Chapter 12).
9. Assess work in progress for the financial year (Chapter 12).
10. Attend site meetings.
11. Preparation presentation and agreement of claims with Employer (Chapter 9).
12. Dealing with extras, claims and contra charges in relation to sub-contract packages.
13. Prepare or agree variations, including Schedule 2 Quotations, with client's surveyor and sub-contractors (Chapter 7).
14. Prepare applications for interim valuations and agree payments with sub-contractors (Chapter 10).

15 Prepare or agree final account with Employer's surveyor and agree sub-contractor final payments (Chapter 11).

The surveyor's legal obligations

The surveyor is normally appointed directly by the client and the two of them enter into a contract of employment. This may be in the form of a letter setting out the essential terms of the contract, such as the services to be provided and the basis of payment, or a standard 'Form of Agreement for the Appointment of a Quantity Surveyor' may be used.

The obligations of the surveyor are governed therefore by the terms of the contract of employment. However, a person who offers professional services has a duty in law to exercise reasonable skill and care in the performance of those services, otherwise he may be liable for damages under the law relating to negligence. His actions are expected to be of an equal standard to those of other qualified persons practising the same profession.

Only a few cases regarding surveyors have come before the courts, and the following are of particular interest.

London School Board v Northcroft Son & Neighbour (1889)

The clients brought an action against Northcrofts, the quantity surveyors, for negligence because of clerical errors in calculations, which resulted in overpayments to the contractor. It was held that as the quantity surveyor had employed a skilled clerk who had carried out a large number of calculations correctly, the quantity surveyor was not liable.

Dudley Corporation v Parsons & Morris (1959)

This case centred around the interpretation of an item in a bill of quantities which stated 'extra over for excavating in rock'. It was unclear whether the item referred to the immediately preceding item of basement excavation or to all the excavation items. The seriousness of the matter from the contractor's point of view was that the item had been grossly underpriced and on remeasurement was some three times the provisional quantity in the bill. It was held by the Court of Appeal that the contractor was entitled only to be paid at the erroneous rate for the total quantity of the item as remeasured.

Tyrer v District Auditor of Monmouthshire (1974) 230 EG 973

The local authority overpaid a contractor because Tyrer, a quantity surveyor who was an employee of the authority, had accepted rates for work, which he must have known were ridiculously high and had also made an arithmetical error when issuing an interim certificate. Tyrer appealed against being surcharged by the District Auditor for the loss sustained by the authority, on the grounds that he was acting in a quasi-judicial position. The appeal was rejected. The quantity surveyor owed a duty to carry out his professional work with a reasonable degree of care and skill.

Sutcliffe v Thackrah (1974) HL 1 Ll Rep 318

Although in this case the defendant was an architect, the quantity surveyor had included in two valuations the value of work, which was defective, not having been advised of the defects by the architect. The client sued the architect for the cost of rectifying the defective work. The House of Lords held that the normal rules of professional negligence applied to all aspects of an architect's duties and, in exercising his functions, he must act impartially. The same obligations also apply to the actions of a quantity surveyor, although it was not a matter under consideration in this case.

John Laing Construction Ltd v County and District Properties Ltd (1982) QB 23 BLR 1

In this case (concerning a contract based on the Standard Form of Building Contract, 1963 edition, July 1977 revision) the quantity surveyor had included in interim certificates fluctuations amounts payable to a sub-contractor, which ought not to have been included because the written notice required in clause 31D(1) had not been given.

The judge concluded that clause 31D(3) of the contract conferred upon the quantity surveyor 'no authority to agree, or to do anything which could have the effect of agreeing, liability as distinct from quantum'. However, he said that 'a quantity surveyor appointed under this or any other contract can as a matter of fact and law be given such authority, or any other authority, by the Employer in any one or more of a number of ways'. The Employer was entitled to have the sums in question excluded from the final account.

Implied obligations

When entering into a contract it is obviously important for a surveyor to be aware of and understand the express terms incorporated into the contract, as failure to comply with these terms may lead to a claim for breach of contract resulting in the award of damages. However it is also important for the surveyor to be aware that other terms may exist within the contract of which he is totally unaware, i.e., implied terms. For example, if a dispute arises in relation to a contract, one of the parties may claim that although a term was not expressly written into the contract its existence is nevertheless implied and if the dispute is taken to the courts a decision will have to be made as to whether such a term was implied in the agreement. Such a situation arose in *Partridge v Morris (1995) CILL 1095*, where a householder engaged an architect to prepare the designs for a house alteration. The architect (Morris) also assisted in the tendering process and recommended a contractor to the householder (Partridge). Unfortunately the recommended contractor was on the verge of insolvency, which affected the quality of the work carried out. The householder terminated the contractor's employment and consequently incurred considerable extra expense in getting the works completed. There was little point in pursuing the original contractor for these costs, as he was insolvent. As a result Partridge sought to recover damages from the architect, part of the claim was that Morris should have reviewed the financial standing of the contractor prior to recommending the acceptance of their tender. Although there was no express term in the agreement between Partridge and Morris requiring the architect to carry out a financial check, the court referred to an RIBA publication, which outlined an architect's duties where it was stated that an architect should discreetly check the financial status of firms providing a tender. Therefore, in this instance the court's decision was that the architect was under an implied duty to research the contractor's financial standing and was in breach of that duty. Surveyors are sometimes asked by their clients to provide advice on the selection and appointment of contractors and in such cases it is important to be aware of what liabilities may be associated with the advice provided.

Obligations to third parties

Through the contract a surveyor will know to whom he is liable, i.e., the other party(s) to the contract. However, since the publication of the Contracts (Rights of Third Parties) Act 1999, which became fully operational on 11 May 2000, it is now possible for third parties to be given beneficial and enforceable rights within a contract. Therefore it is important to be aware when entering into a contract whether third party rights have been incorporated, as this may broaden the surveyor's normal scope of liability. A surveyor also needs to understand that his liability may not be solely limited to his contractual arrangements because, as referred to earlier, there is always the potential for liability to be incurred through the law of tort, and especially the tort of negligence. The well documented case *of Donoghue v Stephenson (1932) AC 562 HL* clearly demonstrated that it was possible to be liable to a third party where they have suffered damage as a result of negligence. By reference to this case it is normally accepted that if an action for negligence against a surveyor is to be successful, a number of issues have to be proven.

1 The surveyor owed the other party a duty of care. This is not always an easy concept to define and it may be useful to review the case of *Donoghue v Stephenson,* where a definition was provided identifying the class of person to whom a duty of care may be owed – 'persons who are so closely affected by my act that I ought reasonably to have them in my contemplation as being so affected when I am directing my mind to the acts or omissions which are called in question'. (Lord Atkin)
2 It has to be shown that the surveyor was in breach of the duty of care.
3 It has to be reasonably foreseeable that injury or damage might arise as a consequence of the breach.
4 It has to be shown that injury, loss or damage has occurred as a result of the breach.

The above points are illustrated in the case of *Smith v Bush (1990) 2 WLR 790.* A surveyor (Bush) was engaged by a building society to carry out a house valuation prior to their issuing a mortgage on the property. The surveyor failed to spot a fundamental structural problem with the property and therefore provided a positive response in his house valuation report, a copy of which was provided to the prospective house purchaser (Smith). Smith purchased the property with a mortgage but shortly the property was damaged because of a structural problem with the chimney construction. Smith had no contract with surveyor, but in this instance the surveyor was found to be liable to the house purchaser through the tort of negligence. It was accepted that the house purchaser relied upon the house valuation report prior to purchasing the property; the surveyor was aware of this and the standard of the report fell below the level normally expected from a professional person.

Surveyors are frequently in a position of providing advice and information in relation to a building project and sometimes this information is provided to parties with whom they have no contract but as demonstrated above, if a surveyor makes a negligent misstatement to a party whom they know will rely upon that statement then there is a potential for being sued for negligence. This can again be illustrated in the case of *J Jarvis v Castle Wharf Developments (2001) 4 Ll Rep 308*, which related to a design and build project within a conservation area for which there were fairly stringent planning requirement. Gleeds, a quantity surveying firm (second defendants), was responsible for providing the tender documentation to the tenderers but the documentation provided did not clearly inform the tenderers that certain aspects of the client's proposal had not been fully agreed with the planning authorities. Jarvis was the successful tenderer but had difficulties in agreeing the proposals with the planning authorities with the end result that an enforcement note was issued to stop the works until Jarvis complied with the planning requirements. Jarvis looked to recover the cost of the delays and revisions from the client and the quantity surveyors. In the first instance Jarvis was successful, even though there was no contract between Jarvis and Gleeds; the court found that Gleeds had been negligent in its representations to the contractor and there had been a breach in its duty of care. The court's decision was overturned on appeal, from further studies of the evidence it became apparent that Jarvis' reliance upon the information provided had in fact been very limited, and this lack of reliance removed one of the cornerstones of bringing a successful action for a negligent misstatement. Despite the initial decision being overturned on appeal, the message is still clear, that is, if a surveying organization negligently provides statements or representations which another party relies upon, they may be sued for economic loss as a result of making a negligent misstatement (*Hedley Byrne v Heller & Partners, 1964*) *1 Ll Rep 485*.

Liability associated with management roles

Since the 1980s many quantity surveyors have looked to diversify from their traditional role and offer clients other services, such as project management and construction management. However, these roles are still relatively new within the construction industry and as a result it is not always certain what services are actually to be provided. Therefore, when a surveyor takes on a management role it is important to clearly define the scope of service to be provided and also be aware of the standards expected from within the industry and from various professional bodies as this may have an impact on what terms may be implied in an agreement between client and surveyor.

The case of *Pozzolanic Lytag v Bryan Hobson Associates (BHA) (1998) 63 ConLR 81* was a dispute relating to a design and build project that was being co-ordinated by a project manager (BHA) whose terms of engagement were based upon a letter sent to Pozzolanic. In the early stages of the project the client had asked BHA to enquire about the contractor's insurance cover. As the project was based on a design and build approach, it was particularly important for the contractor to have professional indemnity cover. The project manager obtained various insurance documents from the contractor and passed them onto the client without checking to see if they were in compliance with the project requirements. Problems arose with the completed project owing to the contractor's defective design. Unfortunately for the client the contractor was insolvent and had no professional indemnity insurance to cover the design inadequacies. The client sought to recover damages from the project manager for failing to ensure that adequate insurance cover was in place. The project manager's defence was that he had fulfilled his duty by obtaining insurance documents from the contractor and passing them onto the client. The court studied, among other publications, the 1992 CIOB Code of Practice for Project Management to determine the scope and nature of the project manager's duties. Within the Code it advised that the typical duties of a project manager included the arrangement of insurance and warranties. Consequently, the court's decision was that BHA was in breach of his duty of care to the client. As project manager he should have checked the insurance particulars, or if he did not possess the necessary knowledge and expertise he should have obtained the appropriate advice, or he should have advised the client to obtain the necessary advice.

A number of surveyors are now involved in construction management where the building works are carried out by trade contractors, who contract directly with the client although their work is organized and supervised by the construction manager. With this arrangement it is usually considered that the risk of delays and problems with work carried out by trade contractors rests with the client and therefore the role of construction manager is often thought to be low risk. However, there still remains a degree of uncertainty as to the full extent of a contract manager's duties and similarly his liability to the client. This uncertainty was demonstrated in the case of *Great Eastern Hotel v John Laing Co Ltd (2005) 99 ConLR 45*, a construction management project for the refurbishment of a hotel. The project did not run smoothly with the end result that it exceeded the budget by nearly 80% and overran by 44 weeks. The client considered that the construction manager (Laing) was responsible for a considerable portion of the delays and extra costs. As a result of the hearing, the court's decision was that Laing had a professional obligation under its agreement with the client to protect the client's interests by giving objective advice and it failed to do so. A further decision about the extent of a construction manager's duties was that Laing was also responsible for the 'scoping' of trade packages. What this means is that when a construction manager organizes a trade package for the client, he is responsible not only for the procurement of the trade contractor but also for ensuring the trade package contains all the necessary works (the scope) for the effective delivery of the project. As a consequence of this case Laing was held to be liable for the client's loss of profits resulting from the delayed opening, the extra cost of carrying out work at a later date through the failure to properly scope the trade contract packages and the cost of the claims presented to the client by trade contractors for delay and disruption for which Laing was held responsible.

Overview

When entering into an agreement it is important to be aware of the express terms as these define the duties and obligations of the parties concerned. It is also important to be aware of the standards of service normally expected through custom and practice, detailed in professional codes of practice or advisory papers and set out in statute or common law as they will have a bearing upon what implied terms may exist within an agreement. It is important to be aware that through the use of collateral warranties or the Contracts (Rights of Third Parties) Act contractual obligations may be owed to third parties. Finally there is always a potential of being found liable to third parties through the tort of negligence.

References

1. NJCC, *Procedure Note No. 2, Alterations to Standard Form of Building Contract* (London: NJCC, 1981).
2. JCT, Series 2, *Practice Note 5* (London: RIBA Publications, 2001), p. 31.
3. JCT, *Standard Building Contract With Quantities, 2005 Edition* (Sweet & Maxwell Ltd), clause 3.12.1.
4. Ibid., clause 2.14.1.
5. Ibid., clause 4.21, 4.23.

Bibliography

1. CECIL, Ray, *Professional Liability* (London: Legal Studies & Services, 1991).

2

Building procurement

The term 'procurement' when used in a building context may be defined as the overall process of acquiring a building. When a client wishes to renovate, extend or construct a new building he will normally need the services of many construction-related organizations to achieve the desired end product. There are a number of alternative procurement methods a client may use to acquire these services. These procurement methods give the client a choice of various management structures, different contractual arrangements and varying degrees of client risk. During the late twentieth century, a number of alternative procurement routes were developed along with alternative options within each procurement routes. As a result it is difficult to classify the various procurement routes into distinct categories, as they sometimes share common attributes. The following descriptions provide a broad appreciation of what procurement routes are potentially available to a client.

Traditional methods

From early in the nineteenth century until about the 1950s, the ways by which building projects were promoted and carried out in the UK conformed to straightforward and well-tried procedures. If the project was small, the building owner (or 'Employer', as he is often called) employed a building contractor to design and construct the building for him. Because buildings generally conformed to a well-defined pattern, contractors had within their organizations the full range of expertise and skills normally required.

In the case of larger projects, the Employer appointed an architect to design the building, and he then produced drawings and a specification. If the architect considered it necessary (and the Employer approved), he then appointed a quantity surveyor to prepare a bill of quantities. Then, on the basis of either the specification and drawings or the bill, contractors were invited to tender in competition to carry out the work. Usually the lowest tenderer was awarded the contract.

Since the mid-1940s, the architect's nomination and/or appointment of the quantity surveyor has been gradually superseded by appointment by the Employer, sometimes before the selection of the architect and, in some cases, the latter's selection is made on the recommendation of the quantity surveyor.

The traditional methods of building procurement are still widely used and their respective distinguishing characteristics are as described below. They have gradually evolved during the twentieth and twenty-first centuries to meet changing circumstances and technological developments, and variants of the main procedures have been introduced from time to time, but the essential principles still apply.

A. *Based on bills of firm quantities*

The building owner commissions an architect to prepare a design and, upon virtual completion of the design, the surveyor prepares a bill of quantities based upon the architect's drawings and specification information. Contractors are invited to price the bill and submit tenders in competition for carrying out the work. The contractor submitting the lowest tender is usually awarded the contract (the JCT Standard Building Contract with Quantities is commonly used).

The essential characteristics of this method are (i) that both the quantities and the unit rates in the bill form part of the contract and (ii) that virtual completion of the design precedes the signing of the contract. Such a contract is a *lump sum contract* (sometimes called a *fixed price contract*) because a price is stated in the contract as payment for the work described in the bill.

Advantages

1 Both parties have a clear picture of the extent of their respective commitments.

2 The unit rates in the bills provide a sound basis for the valuation of any variations to the design.
3 A detailed breakdown of the tender sum is readily available.

Disadvantages
1 The length of time taken in the design of the project and in the preparation of the bills of quantities.
2 The problem of dealing with those variations which are so fundamental or extensive as to change the character of the remainder of the work or the conditions under which it has to be carried out.

B. Based on bills of approximate quantities

This method is largely similar to the preceding one, except that the quantities given in the bill are approximate only and are subject to later adjustment. The essential characteristics are (i) that only the unit rates form part of the contract and (ii) the signing of the contract and the beginning of work on site may proceed before the design is complete.

The bill of quantities is normally specially prepared for the particular project and descriptions of work are as detailed as in a bill of firm quantities, but the time otherwise required for detailed measurement of the quantities is saved, the quantities given being estimates of likely requirements. Sometimes the bill re-uses the quantities which were prepared for an earlier project of a sufficiently similar kind and size.

This method results in a contract (the JCT Standard Building Contract with Approximate Quantities is commonly used) which is sometimes regarded as a lump sum contract although it is not strictly so, there being no total price stated in the contract. In effect, it is very similar to a schedule of rates contract (see D below).

Advantages
1 Construction on site may begin earlier.
2 The extra expense of preparing firm quantities is avoided (although this is offset by the cost of fully measuring the work as actually carried out).

Disadvantages

1 The bills of quantities cannot be relied upon as giving a realistic total cost at tender stage and in consequence, the parties to the contract are less certain of the extent of their commitment.
2 The construction works have to be measured completely as actually carried out, which may prove more costly than to have prepared bills of firm quantities initially.
3 The architect may feel less pressure to make design decisions which ought to be taken at an early stage.

C. Based on drawings and specification

This method closely resembles that described in A above, the difference being that no bills of quantities are supplied to tenderers, who have to prepare their own quantities from the drawings provided. This procedure is intended to be used for relatively small works (say £50,000–£100,000) and for sub-contract works, although it is not unknown for quite large contracts to be tendered for on this basis. A survey carried out for the RICS showed that a small number of contracts let on this basis during 2001 were for a contract value in excess of £10 million.[1]

The essential characteristics are (i) that tenderers are supplied only with complete working drawings and a full specification and (ii) that virtual completion of the design must precede the signing of the contract (the JCT Standard Building Contract without Quantities is commonly used).

Advantages

1 The time required for the preparation of tender documents is reduced, as the time-consuming process of preparing bills of quantities is eliminated.
2 Both parties can have a clear picture of their respective commitments at the time of signing the contract.

Disadvantages

1 No breakdown of the tender sum is immediately available (although the tenderers may be asked to provide a Contract

Sum Analysis, either as a part of their tender submission or subsequently).

2 The valuation of variations presents problems, as indicated above.

3 There is little, if any, control over the percentage rates for additions for overheads and profit to the prime cost of labour, materials and plant elements in dayworks (defined on p. 123). Tenderers are normally asked to state percentage rates to be used in the event of dayworks arising. Where such rates have no effect on the tender sum, there is little incentive to the tenderer to moderate them.

D. Based on a schedule of rates (sometimes known as *measured contracts* or *measurement contracts*)

This method operates in a similar way to that described in B above, tenders being based upon a schedule of rates as explained in (i)–(iii) below. A particular advantage arising from its use is that it allows for a contract to be signed and work to start on site when the design is only in outline form, and in consequence the pre-contract period is reduced considerably.

(i) Standard schedule

A standard schedule lists under appropriate trade headings all the items likely to arise in any construction project, with a unit rate against each item. Standard schedules are produced by both the Property Services Agency of the Department of the Environment[2] and NSR Management.[3] Between them these organizations produce schedules for a variety of works, e.g.:

- Building works
- Mechanical services
- Electrical services
- Painting and decorating
- Landscape management
- Road works.

Tenderers are asked to tender percentage additions (or deductions) to the listed rates, usually by sections or sub-sections, thus allowing for variations in construction costs since the date of the preparation of the schedule used.

Advantage

1 Tenderers using a particular schedule often soon become familiar both with the item descriptions and the rates, and are able to assess percentage adjustments relatively easily.

Disadvantages

1 In comparing and assessing a range of tenders, the surveyor has the task of gauging the effect of a series of variables, making the choice of the most favourable tender difficult.
2 The parties are unable to have a precise indication of their respective commitments.

(ii) 'Ad hoc' *schedule*

This is a schedule specially prepared for a particular project and lists only those items which are appropriate to that project, including any special or unusual items. An '*ad hoc*' schedule may be pre-priced by the surveyor (in which case the form of the tender will be the same as when using a standard schedule) or the rate column may be left blank by the surveyor for the tenderer to insert individual rates against each item. The latter method, because of the absence of quantities, makes the comparison and assessment of tenders much more difficult.

Advantages

1 Tenderers are only required to concern themselves with a restricted range of items, thus enabling them to assess rates or percentages more accurately.
2 Tenderers are able to obtain a clearer picture of the scope of the work from the items listed in the schedule.

Disadvantages

These are similar to those applying to standard schedules.

(iii) Bills of quantities from previous contract

The bill of quantities used will normally be for a comparable type of building of similar constructional form to the proposed project.

It is, in effect, a pre-priced '*ad hoc*' schedule and will be used in the same way.

This is the method of tendering normally used in *serial tendering*, described on p. 73.

Advantages

1 The time required to prepare tender documents is reduced to the minimum.
2 Tenderers have to consider only a restricted range of items.

Disadvantages

1 The parties are unable to have a precise indication of their respective commitments.
2 There may be a considerable discrepancy between the successful tender and the real cost of the work, owing to the approximate nature of the quantities.

E. Based on cost reimbursements (also known as *prime cost* or *cost plus* because the method of payment is reimbursement to the contractor of his prime cost, plus a management fee)

There are three variants of this type of contract, distinguished by the way in which the fee is calculated. Each is dealt with separately below (the JCT Prime Cost Building Contract may be used).

'Prime cost' means the total cost to the contractor of buying materials, goods and components, of using or hiring plant and of employing labour, in order to carry out construction works.

Of all the types of contract, this produces the most uncertainty as to the financial outcome. Tenders contain no total sum and it may be very difficult to form any reliable estimate of the final cost.

It is widely recognized as the most uneconomical type of contract and therefore is one which normally should be used only in circumstances where none of the other types is appropriate. Because of the possibility of inefficiency and waste of resources, contracts of this type need to embody provisions giving the architect some control over the level of labour and plant employed.

One of the attractions of prime cost contracts is that work on site can commence in the early stages of design and this may be all-important to the client. There may be circumstances where, to the client, cost is a less important factor than time. Consequently, a start on site at the earliest possible time may be financially more advantageous in the long run than a lower final cost of construction which might have resulted from the use of another type of contract.

It should be noted that no site measuring is necessary other than as checks on the quantities of materials for which the contractor submits invoices.

The process of calculating and verifying the total prime cost involves a vast amount of investigation and checking of invoices, time sheets, sub-contractors' accounts, etc., which can be both tedious and time-consuming. It is therefore in the interests of the surveyor and the contractor that at the outset a proper system of recording, verifying and valuing the prime cost is agreed and strictly implemented. In addition, it is vitally important to define clearly what is intended to be included as prime cost and what is intended to be covered by the fee. The standard form for the fixed fee variant of this type of contract does so in the appended schedules.[4] The publication *Definition of Prime Cost of Daywork* may be used for this purpose if the standard form is not used, but it will have to be amended as the authors of the publication advise that the definition is designed to be used in conjunction with a building contract and should not be used on its own.

The advantages and disadvantages of prime cost contracts over lump sum and schedule of rates contracts follow.

Advantages

1 The time required for preparation of tender documents and for obtaining tenders is minimized, thus enabling an early start on site to be made.
2 Work on site may proceed before the detailed design is complete.

Disadvantages

1 The parties have the least precise indication of their respective commitments.
2 The cost of construction to the client is likely to be greater than if other types of contract were to be used.
3 The computation and verification of the total prime cost is a long and tedious process.

The variants below differ in the way in which the fee for the contractor's services is determined. Variants (ii) and (iii) are the consequence of a general acknowledgement that it is desirable to provide an incentive to economize in the use of resources on the part of the contractor.

(i) Cost plus percentage fee.

The contractor is paid a fee equal to an agreed percentage of the prime costs of labour, materials and plant used in carrying out the work.

The outstanding disadvantage (to the client) is that the more inefficient the contractor's operations are and the greater the waste of resources, the higher the fee paid to the contractor will be. To counter this, the agreed percentage is sometimes made to vary inversely in comparison to the prime cost, i.e., as the prime cost increases the percentage addition decreases.

The following is a simple illustration of this contractual arrangement, which at the same time will provide a basis for comparison with the fixed fee form. A contract is assumed where the *estimated* total cost of materials, labour and plant was £50,000; the contractor's overheads were calculated by him as 15% and he required 5% for profit. He therefore tendered at 20% overall addition to the prime cost as his fee and his tender was accepted. Assuming that the estimate of prime cost proved at the completion of the contract to be accurate, the total final cost to the client would be:

	£		£
Total prime cost			50,000
15% addition for overheads	7500	20%	
5% addition for profit	2500		10,000
Total cost of contract			60,000

If, however, due to uneconomic organization of the contract, inefficiency and excessive waste, the total prime cost was £55,000 for the same job, then the total cost would be:

	£
Total prime cost	55,000
20% addition for overheads and profit	11,000
Total cost of contract	66,000

LIVERPOOL JOHN MOORES UNIVERSITY
LEARNING SERVICES

As the contractor's overheads chargeable to this job would still be £7500, it follows that the real profit would be £3500. The disincentive to the contractor to work efficiently is thus seen to be strong.

(ii) Cost plus fixed fee.

Under this variant, the fee paid to the contractor is a fixed sum which normally does not vary with the total prime cost, but is based on an estimate of the likely total. The only ground on which the fee might be varied is if either the scope of the work or the conditions of carrying it out were to be materially altered after the contractor tendered.

It should be noted that the fee, if considered in percentage terms, is lower when the prime cost is higher.

Using the same illustration as before, but with the contractor having tendered a fixed fee of £10,000 (made up of £7500 for overheads and £2500 for profit), the sum paid to the contractor as fee would be equivalent to 20% if the total prime cost was £50,000.

If as before, the prime cost was higher for the same reasons, the financial picture would be:

	£	£
Total prime cost		55,000
Overheads	7500	
Profit	2500	10,000
Total cost of contract		65,000

The fee is now equal to 18.18% of the prime cost and the profit portion only 4.55%. If the prime cost rose to £60,000, the fee would equal only 16.67% and the profit portion 4.17%.

(iii) Target cost.

This variant is really one or other of the two preceding ones with another factor added. As an incentive to reducing the total prime cost, the agreement provides for a bonus to be paid to the contractor if the total cost is less than an agreed sum (the 'target') and also a penalty to be paid by him if the total cost exceeds that sum. The bonus and penalty may be a set figure, e.g., 50% of the difference between the total amounts, or any other agreed percentages. Alternatively, there may be a sliding scale of percentages to be used depending on how much below or above the final cost is from the target cost. The target cost is an estimate of the likely total cost.

Taking the same illustration again, it is assumed that the fixed fee method of payment is to be used, and that the agreed sum based on an estimate of likely total cost is £60,000. At the end of the job (assuming the higher cost shown above), the final payment would be calculated like this:

	£
Total prime cost	55,000
Fixed fee	10,000
Total cost of contract	65,000
Deduct penalty, being 50% of excess over £60,000	2,500
Amount of final payment	62,500

In effect, the fee has been reduced to £7500, which is only just enough to meet overheads, leaving nothing for profit. It follows therefore that the contractor will need to be very satisfied that the 'agreed sum' is a realistic estimate of likely cost and the surveyor must give him every reasonable opportunity and assistance to satisfy himself in that regard.

However, if the contractor were able to reduce the prime cost, say to £48,000, final payment would be calculated as:

	£
Total prime cost	48,000
Fixed fee	10,000
Total	58,000
Add bonus, being 50% of saving over £60,000	1,000
Amount of final payment	59,000

The fee has now increased to £11,000, including £3500 for profit, the overheads remaining at £7500. As a percentage, the profit is 7.29% of the prime cost. Thus, under this contractual arrangement, an extra cost or saving is shared between the client and the contractor.

The composite nature of contracts

Although, for convenience, classifications and type labels such as the foregoing are commonly used, in practice contracts often combine the characteristics of two or more types. McCanlis has pointed out[6] that, for example, a lump sum contract based on bills of firm quantities often contains items with provisional quantities requiring re-measurement and therefore such items bear the characteristics of a schedule of rates contract. Also, the provisional sums included in the bills for daywork and expended on work, which is not readily measurable or not reasonably priceable as measured work, form a prime cost plus percentage fee contract within the main lump sum contract. Thus, many contracts which are regarded as of the lump sum type are, in reality, a combination of several types. Other examples of the composite nature of many contracts are given in the reference above.

Circumstances in which the various types of contract may be used

Usually, the circumstances peculiar to a project will indicate which type of contract is most appropriate. Occasionally, more than one type might be suitable, in which case, the one which seems to offer the greatest benefits to the client should be chosen, as the client is the one who will be paying the bill.

The suitability of the types of contract discussed earlier may be related to varying circumstances as follows.

Based on firm bills of quantities

(i) When there is time to prepare a sufficiently complete design to enable accurate quantities to be measured.
(ii) When the client's total commitment must be known beforehand, for example, in order to make adequate borrowing arrangements or when approval to the proposed expenditure has to be obtained from a finance or housing committee of a local authority, from a central government department or from a board of directors.

Based on bills of approximate quantities

(i) When the design is fairly well advanced, but there is insufficient time to take off accurate quantities or the design will not be sufficiently complete soon enough for that to be done.

(ii) When it is desired to have the advantages of detailed bills but without the cost in terms of time and/or money.

Based on drawings and specification

(i) When the project is fairly small, i.e., up to a value of, say, £50,000–£100,000.
(ii) Where time is short and the client considers it to be less important to have the benefits of bills of quantities than the early completion of the construction work, while retaining the advantages of a lump sum contract.

Based on schedule of rates

(i) When the details of the design have not yet been worked out or there is considerable uncertainty with regard to them.
(ii) When time is pressing.
(iii) When *term contracts* are envisaged. These are appropriate where a limited range of repetitive work is required to be carried out, such as the external redecoration of an estate of houses. The contractor tenders on the basis of unit rates, which are to remain current for the stated 'term', usually one year or the estimated period, which the proposed work will take.

Based on prime cost plus fee

(i) When time is short and cost is not as important as time.
(ii) When the client wishes to use a contractor who has worked satisfactorily for him before and whom he can trust to operate efficiently, while being prepared to pay the higher cost entailed in return for the advantages.
(iii) In cases of emergency, such as repairs to dangerous structures.
(iv) For maintenance contracts.
(v) For alterations jobs where there is insufficient time or it is impracticable to produce the necessary documentation.

The following examples will illustrate some of the foregoing considerations.

Example 1

A client company owned a department store in a prime position in Manchester. It had acquired the adjoining site for an extension, which it was estimated would cost £500,000. The Board of Directors wanted the work to be done by a particular contractor who had done work for them before.

Apart from the choice of the contractor, the important consideration from the client's point of view was the need to commence trading in the new extension at the earliest possible date and also to restrict the disturbance to their normal trading operations to as short a period as possible.

Of the variants to the lump sum type of contract, only the drawings and specification type could be considered on the ground of time, but because of the early start requirement it was rejected. As it was thought unlikely that there would be many variations during the course of construction, the schedule of rates type was rejected too.

The clients were not so concerned about cost as about time, and because they knew the contractors and were satisfied that they would do an efficient and speedy job, the type of contract eventually recommended (and used) was the target cost form of prime cost plus a fixed fee. The contractors produced an estimate cost of £530,000 and after some negotiation the target cost was finally agreed at £518,000.

Example 2

The clients were manufacturers of clothing and had a new factory in the early stages of construction in a suburb of Birmingham. The contract was based on bills of quantities and the work was proceeding on schedule. Another site had been acquired by the clients for a building of similar size and form of construction about 18 miles away at Coventry. The clients were anxious to begin production as soon as possible in both factories.

As the two sites were reasonably close and time was an important factor, the type of contract based on a pre-priced schedule of rates (namely, the priced bills of quantities for the first contract) seemed the best choice. The prices in those bills could only be used, of course, if the same contractor was able and willing to take on the second contract as he held the copyright in the prices.

In the event, he was prepared to do so and percentage additions to each of the work sections in the bills were negotiated between the surveyor and the contractor's estimator. A new 'Preliminaries' section of the bills appropriate to the second project was prepared and prices negotiated for the items in it.

Example 3

An Inner London Borough Council was proposing to improve some 1500 of its older dwellings by replacing all the sanitary appliances with modern equipment over the next three or four years. Finance had already been allocated for the work to commence in the current financial year and a start was to be made as soon as contracts could be arranged.

The main characteristic of this situation was the large amount of repetition of relatively few items of work. Time was also a factor in so far that an early start was essential if allocated finance during the current financial year was to be fully used.

The circumstances all pointed to 'term contracts', and, accordingly, contracts based on *ad hoc* schedules were used. Schedules of the plumbing and associated builder's work items were drawn for tenderers to insert their own rates against the items. Provision was made in the contracts for the rates to be updated at 12-monthly intervals.

The risk factor

An important element of construction contracting is risk, i.e., the risk the parties take that the implementation of the contract may be detrimental in some degree to their financial or other interests. On his part, the contractor takes the risk that his anticipated profit will be reduced or converted into a loss as a result of the outworking of the conditions under which the contract is carried out. On his part, the Employer takes the risk that he will become liable for a greater total cost than he envisaged when initiating the project.

The types of contract described above carry a varying degree of risk for the parties. Generally speaking, the contractor's risk increases as the Employer's risk reduces and *vice versa*. Thus, the contractor bears a high degree of risk where the contract is based on drawings and specification only, for two reasons. First, as it is a lump sum contract, he must estimate his expected costs as accurately as possible because any adverse mistake will reduce his profit. Second, there being no bill of quantities provided for tendering purposes, he must take off his own quantities from the drawings in order to formulate his tender, and, again, any error he may make in the process will affect his profit margin.

The Employer's risk is small in the situation just described. He knows at the outset what his financial liability will be and is under no contractual obligation to reimburse the contractor for any errors which he may have made in preparing his tender.

At the other end of the spectrum, where a cost reimbursement contract is used, the contractor's risk is reduced considerably because he is paid his full costs and a fee in addition. His only risk is in pitching at the right level the fee which he tenders but, even so, he is most unlikely to make a loss. On his part, however, the Employer bears the risk of the prime cost becoming much higher than estimated, owing perhaps to an inefficient site agent or to wastage of resources.

Other types of contract fall within these two extremes, and Figure 2.1 indicates a ranking of them according to the allocation of risk borne by the parties.

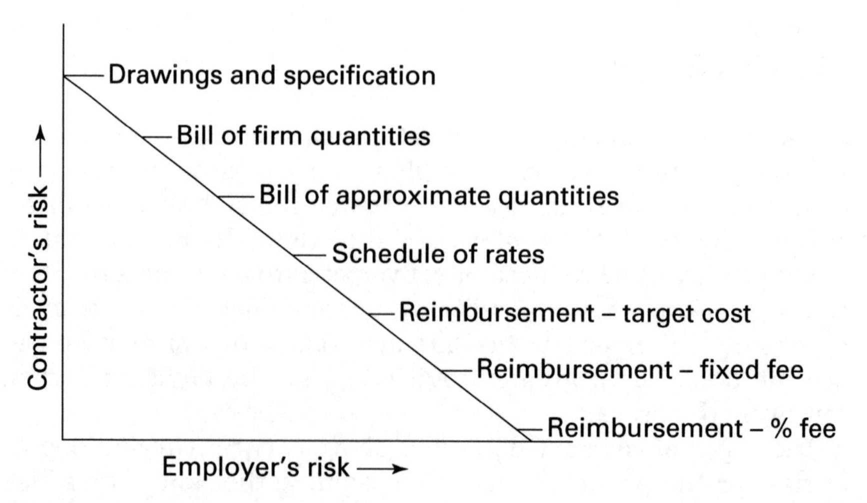

Figure 2.1 Allocation of risk

Relationships between the parties

In all the traditional types of contract, the pattern of relationships between all the parties is normally similar. Figure 2.2 illustrates these relationships and indicates the lines of communication between them.

Contractual links exist between

(i) the Employer and each of his professional advisers (architect, quantity surveyor and consultant engineers);
(ii) Employer and contractor;
(iii) the contractor and each sub-contractor.

There may also be Warranty Agreements between the Employer and sub-contractors if the Employer wishes (as, for example, by use of Form SCWa/E in conjunction with the JCT Standard Building Contract).

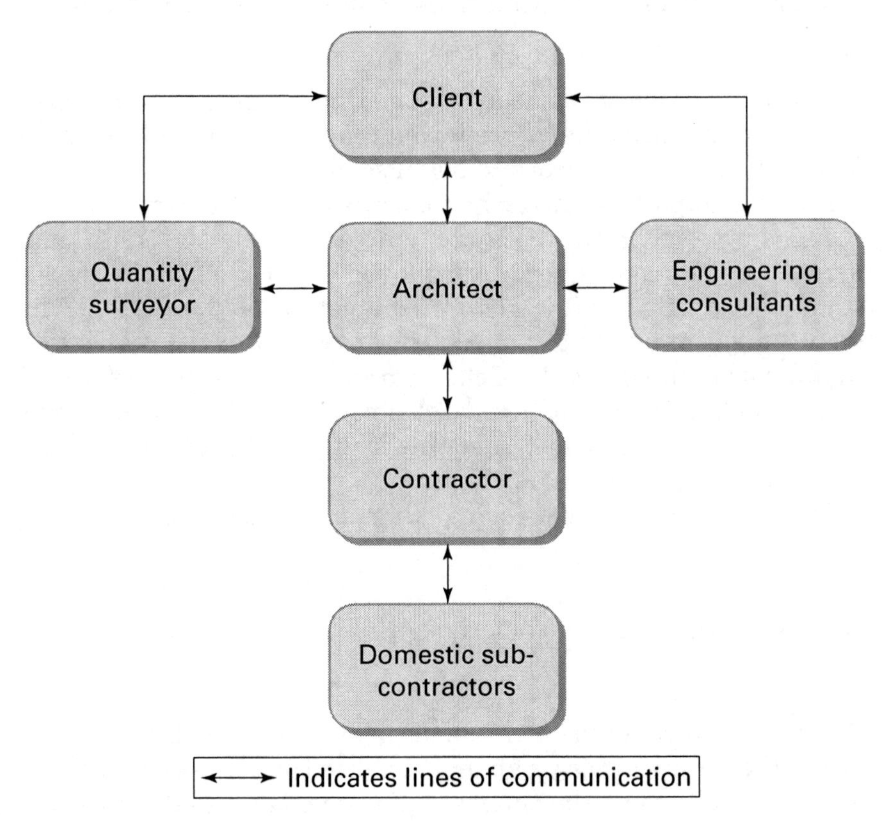

Figure 2.2 Relationships and communication links

Alternative methods

Since the early 1960s, various alternative ways of promoting and carrying out construction projects have been devised with varying degrees of success. The reasons why many building owners and developers considered that the traditional procedures were no longer satisfactory were:

(i) The rapidly spiralling cost of construction meant that large sums of money had to be borrowed to finance projects;
(ii) high interest rates meant that the time occupied by the traditional procedures resulted in substantial additions to the construction cost;
(iii) clients were becoming more knowledgeable on construction matters and were demanding better value for money and an earlier return on their investment;
(iv) high technology installations required a higher quality of construction.

Attention has focused on reducing the time traditionally occupied in producing a design and preparing tender documentation, thus enabling construction work to begin sooner.

Another important factor has been bringing the contractor in at an early stage in the design of a project. Under the traditional procedures, the contractor rarely played any part until the tender stage was reached, after virtual completion of the design. The increasing complexity of projects, however, led to the realization that it was in the interests of clients and architects to use the vast amount of knowledge and practical experience of contractors early in the design process, and that this would make a valuable contribution to a successful outcome.

The following are the principal alternative methods of building procurement now in use.

A. Design and build (sometimes called *design and construct* or *package deals*)

When first introduced in its modern form, this method usually meant the contractor using an industrialized building system which was adapted to meet the client's requirements – usually straightforward, rectangular warehouse or factory buildings with a minimum of design. Nowadays the method is applied to a wide range of building types.

Essentially, the contractor is responsible for the design, for the planning, organization and control of the construction and for generally satisfying the client's requirements, and offers his service for an inclusive sum. A proprietary, prefabricated building system may or may not be used.

The procedure is initiated by the client (or an architect on his behalf) preparing his requirements in as much or as little detail as he thinks fit. These are then sent to a selection of suitable contractors, each of whom prepares his proposals on design, time and cost, which he submits together with an analysis of his tender sum. The client then accepts the proposals, he is satisfied best meet his requirements and enters into a contract (e.g., the JCT Design and Build Contract is a suitable form of contract) with the successful tenderer. The latter then proceeds to develop his design proposals and to carry out and complete the works.

The client may use the services of an independent architect and quantity surveyor to advise him on the contractor's proposals as to design and construction methods and as to the financial aspects, respectively. He may also appoint an agent to supervise the works and generally to act on his behalf to ensure that the contractor's proposals are complied with.

Figure 2.3 shows the normal pattern of relationships and indicates the lines of communication between the parties in a design and build contract.

Contractual links exist between

(i) the client and the contractor;
(ii) the contractor and each sub-contractor;
(iii) the client and each of his independent advisers.

Advantages

1 Single point responsibility is provided, i.e., the contractor is solely responsible for failure in the design and/or the construction.
2 The client has only one person to deal with, namely, the contractor, whose design team includes architects, quantity surveyors, structural engineers, etc.
3 The client is aware of his total financial commitment from the outset.

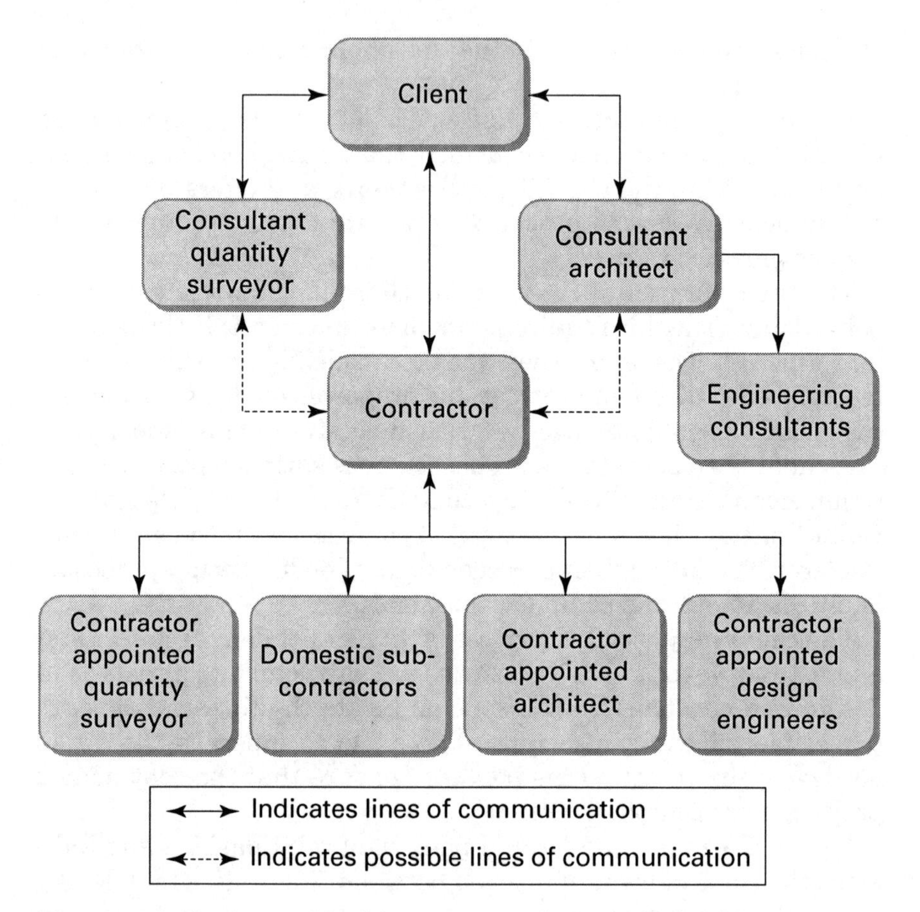

Figure 2.3 Design and build contracts: relationships of the parties

4 Close intercommunication between the contractor's design and construction teams promotes co-operation in achieving smoother running of the contract and prompt resolution of site problems.

Disadvantages

1 Variations from the original design are discouraged by the contractor and, if allowed, are likely to be expensive.
2 The client has no means of knowing whether he is getting value for money unless he employs his own independent advisers, which adds to his costs.
3 If the contractor's organization is relatively small, he is unlikely to be as expert on design as he is on construction, and the resulting building may be aesthetically less acceptable.

B. Management fee

The management fee procurement route is based upon the client employing a professional team to advise him on design and cost issues and in addition a management consultant to advise on and supervise the construction aspects of the project. The construction work is broken down into individual packages and let out to sub-contractors throughout the progress of the project as and when design details are finalized. This approach enjoyed a degree of popularity from the early 1970s to the end of the 1980s. A survey of usage of various forms and types of contract carried out in 2001[7] showed that the use of management fee contracts peaked in 1991 when 27.3% of projects (by total value) used this form of procurement, but by 2001 this figure had fallen to 11.9%. This fall in usage is possibly a reflection of some client dissatisfaction with this procurement route and to some extent a change in the economic climate from the 1980s. There are two main variations of the management fee approach as identified below.

Management contracting

The principal characteristic of management contracting is that the management contractor does none of the construction work himself but it is divided up into work packages which are let to works contractors, each of whom enters into a contract with the management contractor. The latter is normally either nominated by the client on the basis of the contractor's previous experience of management contracting or is selected by competition based upon tenders obtained from a number of suitable contractors for (a) the management fees and (b) prices for any additional services to be provided before or during the construction period (unless to be paid for on a prime cost basis, in which case the tenders will include percentage additions required to the respective categories of prime cost). The successful contractor will then enter into a contract with the client (e.g., the JCT Management Building Contract).

The management contractor's role therefore is that of providing construction management service on a fee basis as part of the client's management team – organizing, co-ordinating, supervising and managing the construction works in co-operation with the client's other professional consultants. As part of his service, he provides and maintains all the necessary site facilities, such as offices, storage and mess huts, power supplies and other site services, common construction plant, welfare, essential attendances on the works contractors (e.g., unloading and storing materials, providing temporary roads and hardstandings and removing rubbish and debris) and dealing with labour relations matters.

Figure 2.4 shows the usual pattern of relationships and indicates the lines of communication between the parties to a management contract.

Contractual links exist between

(i) the client and each member of the design and management team (including the management contractor);
(ii) the management contractor and each of the works contractors.

Advantages

1 Work can begin on site as soon as the first one or two works packages have been designed.

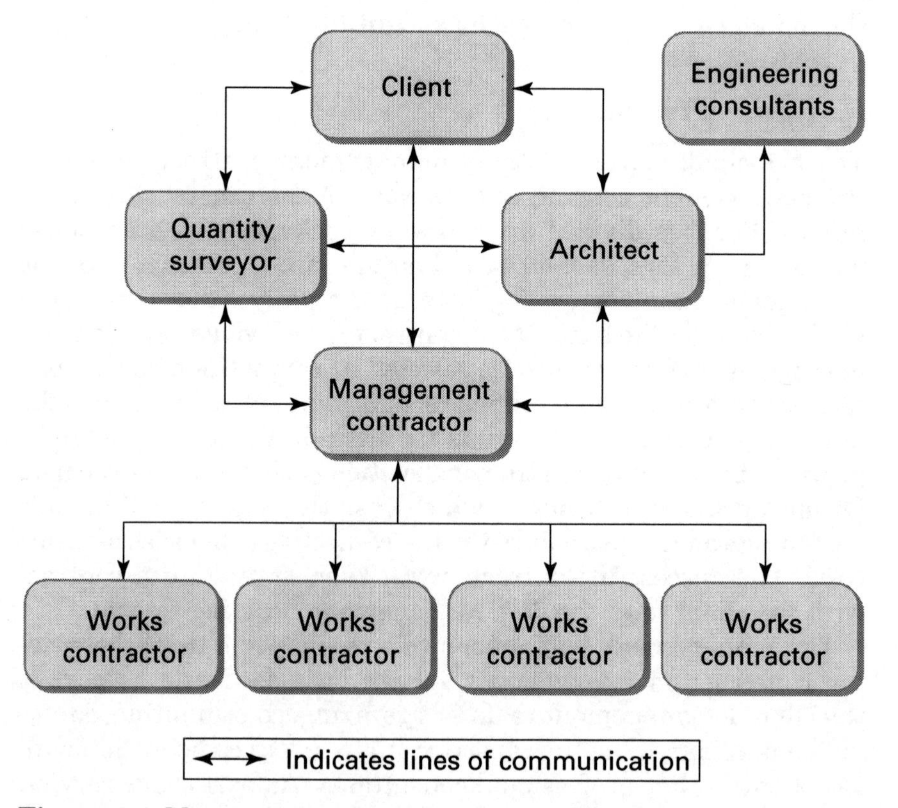

Figure 2.4 Management contracts: relationships of the parties

2 Overlapping of design and construction can significantly reduce the time requirement, resulting in an earlier return on the client's investment.
3 The contractor's practical knowledge and management expertise are available to assist the design team.
4 Where the nature and extent of the work may be uncertain, as in refurbishment contracts, the design of later work packages may be delayed until more information becomes available as the work progresses, without extending the construction period.
5 The contractor, being part of the client's team, is able to identify with the client's needs and interests.
6 Because works contracts are entered into close to the time of their commencement on site, they can be based on firm price tenders.

Disadvantages

1 Uncertainty as to the final cost of the project until the last works contract has been signed.
2 The number of variations and the amount of re-measurement required may be greater than on traditional contracts because of the greater opportunity to make changes in design during the construction period, because of problems connected with the interface between packages, and because packages are sometimes let on less than complete design information.

Construction management

Construction management is similar to management contracting to the extent that it is a professional consultant service to the client, provided on a fee basis, with the design and construction services being provided by other organizations.

The construction manager is responsible for the organization and planning of the construction work on site and for arranging for it to be carried out in the most efficient manner. The construction work itself is carried out by a number of contractors, each of whom is responsible for a defined trade package. All the trade packages together constitute the total project. Each trade contractor enters into a direct contract with the client.

The construction manager's duties normally include any or all of the following:

(i) co-operation and consultation with the other members of the client's professional team;
(ii) preparation and updating of a detailed construction programme;
(iii) preparation of materials and components flows and arranging for advance ordering;
(iv) determining what site facilities and services are required and their location;
(v) breaking down the project into suitable trade packages in consultation with the other members of the client's team and recommending suitable contractors to be invited to tender for trade packages;
(vi) obtaining tenders from contractors and suppliers;
(vii) evaluating tenders and making recommendations on them to the client's team;
(viii) co-ordinating the work of the trade contractors to ensure that it is carried out in accordance with the master programme;
(ix) establishing all necessary management personnel on site with responsibility to manage and supervise the project;
(x) dealing with any necessary variations to the work, providing the design manager with estimates of their likely cost and subsequently issuing instructions to trade contractors;
(xi) submitting to the quantity surveyor applications from trade contractors for periodic payments and all necessary documentation enabling the final accounts of trade contractors to be settled.

Figure 2.5 shows the normal pattern of relationships between the parties and indicates the lines of communication between them.

Contractual links exist between

(i) the client and the construction manager;
(ii) the client and each of the trade contractors;
(iii) the client and each of his professional advisers (architect, quantity surveyor, engineering consultants, etc.).

Advantages

1 The construction work is more closely integrated into the management of the project.

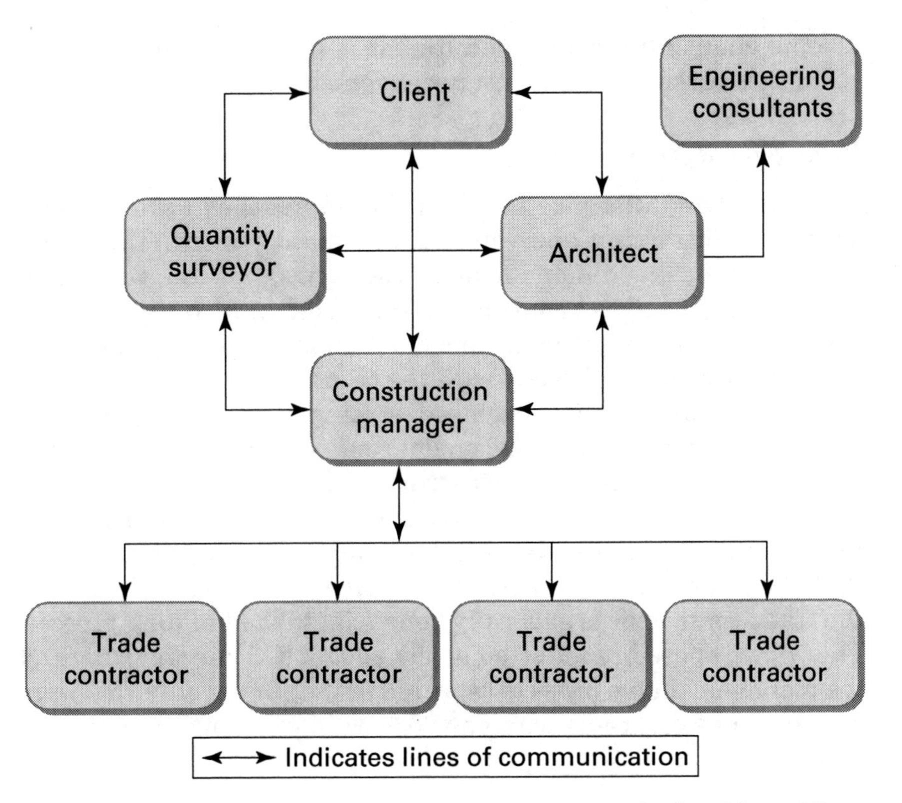

Figure 2.5 Construction management contracts: relationships of the parties.
Note: Lines of communication also exist between the architect and quantity surveyor and each of the trade contractors. To avoid confusion, these lines are omitted.

2 Close liaison between the construction manager and design manager leads to prompt identification of and decisions relating to practical problems.
3 Detailed design can continue in parallel with construction, trade packages being let in succession as the design of each is completed, thus shortening the project time.
4 Privity of contract between the client and each of the trade contractors provides the client with a readier means of redress in the event of difficulties, such as delays, arising.

Disadvantages

1 The client has one more consultant and a number of contractors with whom to deal instead of only one main contractor.

2 The client's financial commitment is uncertain until the last of the works contracts has been signed.

Management fee overview

There is little difference between the approaches used in management contracting and construction management. The major difference is that in management contracting all the works contractors are in a direct contractual relationship with the management contractor, whereas in construction management the trade contractors contract directly with the client. Over a period of time, a general opinion has developed stating that for experienced clients the construction management route is possibly the preferred route of the two. Although construction management causes the client much more administration and possibly a greater level of risk by contracting directly with the trade contractors, it is claimed that the closer arrangement and involvement of the client with the trade contractor is generally beneficial to the building process. This view is possibly reflected in the 2001 RICS survey[8] where, of the management fee projects captured in the survey, only 15% were management contracts whereas 85% were construction management contracts (percentage by value).

C. Government procurement routes

During the late twentieth and early twenty-first centuries the government was looking at ways in which it could improve the procurement of construction works within the public sector while at the same time encouraging greater efficiencies within the industry as a whole. In April 2000, government policy had identified three recommended procurement routes: Design and Build, Private Finance Initiative (PFI) and Prime Contracting. It was recommended that traditional procurement routes should only be considered where it could be clearly demonstrated that they provided better value than the three recommended routes.[9]

Private Finance Initiative (PFI)

A definition of PFI is 'Where the public sector contracts to purchase quality services, with defined outputs from the private sector on a long-term basis, and including maintaining or constructing the necessary infrastructure so as to take advantage of private sector management skills incentivized by having private finance at risk'.[10] Although PFI is one of the preferred procurement routes for government works there are limits as to its suitability. Its use is recommended where it offers clear value for money, and is considered to be more appropriate for large projects, i.e., in excess of £20m, and projects that have significant ongoing maintenance requirements.[11] To illustrate the limited use of PFI the government estimated that it would account for only 11% of its total investment for 2003–2004.[12]

There are a number of ways in which PFI may operate but perhaps the most common is 'build-own-operate-transfer' (BOOT). In this situation contractors would be provided with details of the client's requirements, regarding function, facilities, maintenance, etc. and be asked to tender for the scheme. The successful tenderer (often a consortium of companies) would be responsible for the design and construction work, and on completion it would continue to be responsible for the running and maintenance of the works. This responsibility may run for a set period of time, e.g., 20–30 years where upon the works revert back to the government or local authority. Also within the PFI agreement is a payment mechanism to allow the PFI contractor to be reimbursed all its construction, finance and maintenance costs. The payments will not normally commence until the work is finished and is operational. Obviously if the PFI contractor performs inefficiently then the payments may not be sufficient to cover its costs and generate any profit, and that is the risk the contractor takes. Since its inception PFI has been used to provide numerous facilities: schools, hospitals, leisure centres, bridges, roads, tram systems and rail links, e.g., the Channel Tunnel Rail Link was a PFI project valued at £4178m. There are conflicting views as to the benefits of PFI but from a report prepared by the National Audit Office they found that PFI has been able to deliver built assets on time and within the expected price.[13]

Prime contracting

Prime contracting has developed to a large extent as a result of reports produced in the 1990s (Constructing the Team, Sir Michael Latham and Rethinking Construction, Sir John Egan), which encouraged the use of teamwork and the integration of the whole supply chain into the construction process. The general principle is that the work will be undertaken by a prime contractor who will be responsible for bringing together, co-ordinating and managing the supply chain (i.e., all the organizations necessary for carrying out the works – designers, engineers, sub-contractors, suppliers, etc.). A prime contractor does not have to be a contracting organization, the term is used to identify any organization that is capable of providing the above service, therefore the role may be taken by a designer, project manager, contractor, etc. The claimed advantage of the system is that the prime contractor will provide a single point responsibility and should bring to the client's project a well-established supply chain that has a proven working relationship, an ability to contribute to all aspects of the project in a harmonious relationship which should lead to a reduction in the client's capital and maintenance costs. The system is designed to overcome the traditional fragmented nature of the UK construction industry and seeks to remove the adversarial approach adopted on many projects.

Prime contracting is not just concerned with the initial provision of a building, there is also a general expectation that the prime contractor must be able to demonstrate during the early occupation of the project that the cost and performance requirements can be met[14] which may well require the prime contractor to employ value management techniques. It is also recommended that consideration should be given to letting the contract on a target cost basis with the prime contractor (and the whole supply team) having a financial share in any savings that may be made but with them also being liable for a share of any overspend incurred.

E. Project management

Project management is not a procurement system in itself, in that it does not include the site construction process but only its general supervision. It has been defined as 'the overall planning, control and co-ordination of a project from inception to completion aimed at meeting a Client's requirements in order that the project will be completed on time within authorized cost and to the required quality standards'.[15]

A number of project management organizations came into existence during the late 1970s and 1980s to meet the need for the management of projects, which were becoming larger and increasingly complex. Many clients do not have the in-house skills and experience necessary for the successful management of construction projects and so need to employ an independent project management company to do it for them.

Quantity surveyors, by their training and experience in financial and contractual matters, coupled with a detailed knowledge of construction processes, are well qualified to offer a project management service, and a number of established quantity surveying practices have set up associated companies offering such a service. Other groups of professionals, such as architects, engineers, and building and valuation surveyors, are also now filling this role.

The project manager, in effect, becomes the client's representative, with authority to supervise and control the entire planning and building operation from acquisition of the site to completion of the project and settlement of the accounts. The service he provides is essentially one of planning, organizing and co-ordinating the services provided by surveyors and lawyers in relation to site acquisition; the architect, engineers and quantity surveyor in relation to project planning and design; and the contractor and subcontractors in carrying out the site construction work; but does not include the carrying out of any of their duties himself.

Figure 2.6 shows the normal pattern of relationships between the parties and indicates the lines of communication between them.

Contractual links exist between

(i) client and project manager;
(ii) client and each of his professional advisers;
(iii) client and contractor;
(iv) contractor and each sub-contractor.

F. Partnering

The use of partnering was developed by a number of state authorities in the US during the 1980s and met with a degree of success, and as a consequence during the 1990s the idea of partnering came to prominence in the UK. Partnering is similar to project management in the fact that it cannot really be categorized as a procurement route but unlike project management it does not create a management structure for the delivery of the client's project. In its simplest form partnering is a philosophy of introducing trust and

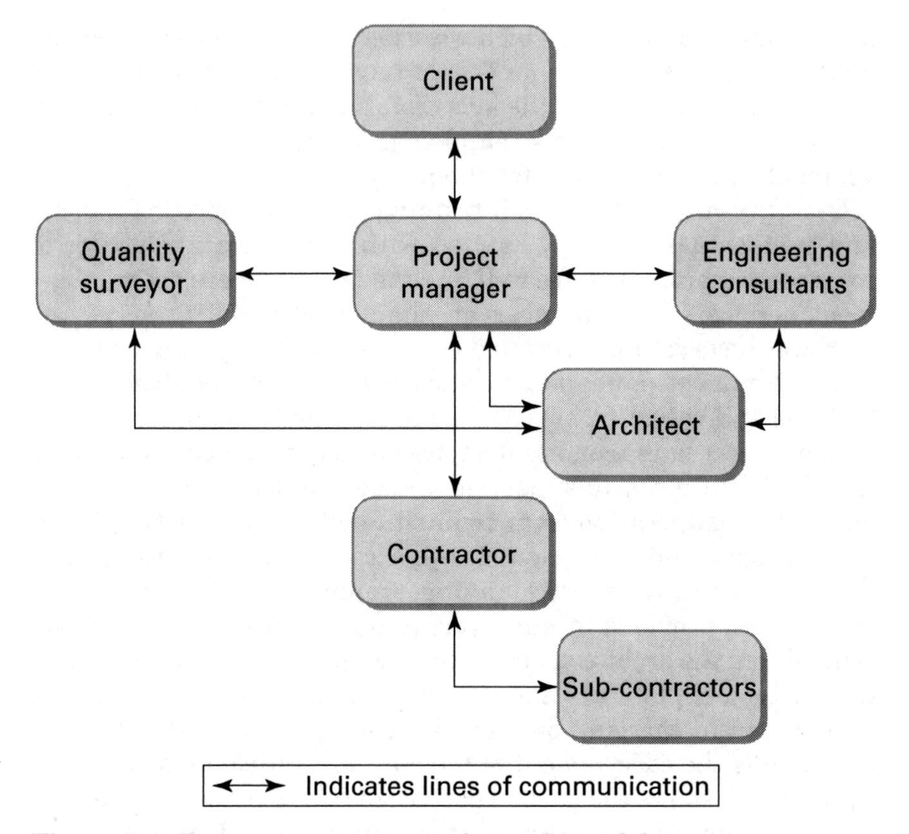

Figure 2.6 Project management: relationships of the parties

teamwork into the construction process, engendering commitment to common objectives, encouraging a shared focus on the project and on how it may best be delivered, seeking solutions to problems and removing adversarial attitudes. The principle of partnering should extend through the whole supply chain, it should be seen as an alliance between client, consultants, contractor and sub-contractors. If a contractor fails to encourage the same principles between himself and his suppliers and sub-contractors then the client is unlikely to reap the full rewards of the system.

Where partnering is to be used on a project it is recommended that a project agreement is drawn up that will identify the partnering principles to be employed throughout the project and signify the parties' commitment to those principles. The problem with some early projects where partnering was tried was that the parties involved were not totally clear on how the process should operate and there were reports of early projects eventually reverting back to the old adversarial approach. As a consequence it was claimed that partnering could not really exist on trust alone, and what was needed was a more rigid structure at the outset which has led to the publication of at least three standard forms of agreement for partnering, PPC 2000 produced by the Association of Consultant Architects, the NEC Partnering Agreement (Option X12) and the Public Sector Partnering Contract (PSPC) developed by Knowles Management and the Federation of Property Societies. PPC 2000 is a contract for the project, which all the partnering members are required to sign. The contract contains express terms on how the partnering approach is to operate as well as all the normal terms relating to the administration of a project, e.g., payment, insurance, variations and extensions of time. The NEC and PSPC partnering arrangements adopt a different approach; they are in fact optional documents that may be incorporated into standard forms of contract, as such they are much shorter documents as the bulk of their content is based solely around the partnering principles.

Framework agreements

The advantages of partnering is that it should put the parties in a win/win situation, the site construction process should run more smoothly, there is scope for increased innovation and improvement in quality, less waste in time and resources and overall better value for money. However, these advantages may not be quite so evident in one-off projects and it is claimed that a client with an ongoing programme of construction works is more likely to benefit from the accumulation of partnering benefits that may be acquired through strategic alliances. Similar arguments can also be applied to prime contracting and as a result a number of clients (e.g., BAA, NHS, the Prison Service) with a large and continuous construction programme have developed the idea of producing framework agreements. The principle involved is that a number of key organizations or consortiums within the construction industry (architects, engineers, surveyors, contractors) are put through a vigorous pre-qualification process which if successfully concluded results in them becoming the client's preferred bidders in future projects. Thus, the framework agreement allows for more continuity within the parties and allows the partnering benefits to grow and accumulate. The JCT have produced two standard forms of framework agreement one binding (FA) and one non-binding (FA/N).

G. Joint venture

This is an arrangement for building procurement which has developed out of the increasing complexity of construction projects, often with unorthodox methods or sequences of construction. It is applicable to large-scale projects of, say, £5m value or more, where there is a higher than normal proportion of engineering and other specialist services. It has been defined as follows:

> A partnership between two or more companies covering building, mechanical and electrical engineering, or other specialist services for the purpose of tendering for, and executing a building or civil engineering contract, each of the participating companies having joint and several liability for their contractual obligations to the employer.[16]

The basis of such projects is a single contract to which all the partner companies to the joint venture are signatories. Any of the standard forms of contract in current use may form the contract document, and in the publication just quoted the NJCC recommended the use of JCT standard forms for projects not primarily of an engineering nature. Modifications will need to be made to cover joint and several liability of the participating companies and to provide for any performance warranty which the Employer may require.

Joint venture has the effect of making specialist contractors, who under more orthodox arrangements would be sub-contractors, partners with the building contractor. This does not preclude the use of sub-contractors (whether, named or listed), a role which is filled by secondary specialist contractors who are not partners in the joint venture.

Advantages

1 The combined resources of participating companies result in greater economy in the use of specialist manpower, design and equipment.

2 There is a greater degree of unified action between the contracting parties.
3 Lines of communication between the Employer's professional advisers and the building contractor and specialist contractors are shorter and better defined.
4 Improved integration of work sequences results in a shorter contract time and fewer management problems.

Disadvantages

1 If one JV partner withdraws, the remaining partner or partners must accept total liability to complete the project.
2 A separate, unified JV bank account must be arranged and maintained by all the partners.
3 Separate insurance policies must be taken out on behalf of all the JV partners, specifically for the proposed project.
4 A board of management with a single managing director needs to be appointed jointly by the JV partners to ensure unified actions.

The pattern of relationships in a Joint Venture project and the lines of communication between the parties are similar to those illustrated in Figure 2.2, but with the JV partners together fulfilling the role of 'Contractor'.

Contractual links exist between

(i) the JV participating companies;
(ii) the Employer and the JV partners collectively;
(iii) the Employer and each of his professional advisers;
(iv) the JV partners and each of the sub-contractors.

Bibliography

1. CLAMP, H., and COX, S., *Which Contract?* (London: RIBA Publications, 1989).
2. CHAPPELL, David, *Which Form of Building Contract?* (London: ADT Press, 1991).
3. TURNER, Dennis, *Building Contracts: A Practical Guide*, 5th edition (London: Longman, 1994).

References

1. DAVIS, LANGDON and EVEREST, *Contracts in Use* (London: RICS, 2003).
2. *PSA Schedule of Rates for Building Works, 2005* (London: HMSO, 1985).
3. *National Schedule of Rates for Building* (Aylesbury: NSR Management, 2005).
4. *Prime Cost Building Contract* (London: Sweet & Maxwell Ltd, 2005).
5. *Definition of Prime Cost of Daywork Carried out under a Building Contract*, 2nd edition (London: RICS Books, 1975).
6. McCANLIS, E. W., *Tendering Procedures and Contractual Arrangements* (London: RICS Books, 1967), p. 10.
7. *Contracts in Use*, op. cit.
8. ibid.
9. *Achieving Excellence in construction Procurement Guide 06: Procurement and Contract Strategies* (London: Office of Government Commerce, 2003), p.4.
10. ibid.
11. *PFI: Meeting the Investment Challenge* (London: HM Treasury, 2003), p 2.
12. ibid.
13. *PFI: Construction Performance* (London: National Audit Office, 2003), p. 6.
14. Achi*eving Excellence in Construction Procurement Guide 06*, op. cit., p. 10.
15. *Code of Practice for Project Management for Construction and Development* (Englemere: CIOB, 1992).
16. NJCC, *Guidance Note 1 – Joint Venture Tendering for Contracts in the UK* (London: RIBA Publications Ltd, 1996), p. 1.

3

Contract selection

Introduction

It is normal, but not essential, for standard forms of contract to be used as the basis for an agreement between a developer and a contractor. The complex nature of construction, together with the high levels of risk often involved, usually mean that the use of a standard form of contract is advisable, certainly for all but the smallest of projects. Standard forms of contract offer the advantages of being carefully drafted by experts, being tried and tested and therefore 'known quantities', and being recognized by both the industry and the courts. Although such forms may have their individual weaknesses, it is often safer to rely on the 'devil you know'!

Surveyors must have a working knowledge of the availability and suitability of the various standard forms of construction contracts. Familiarity with one standard form of contract should not be a reason for surveyors always recommending the use of that form for all the projects they are involved with. It is imperative, if clients' interests are to be safeguarded, to recommend the form of contract most suited to each set of project circumstances. In this respect, it is incorrect to suggest that one form of contract is always better than another – the various standard forms have been drafted for use in specific circumstances. Using an inappropriate standard form of contract for a project is dangerous. It will often mean that objectives, in terms of time, cost and quality, are not fully realized and that the likelihood of disputes will increase. It is therefore essential for surveyors to develop effective contract selection skills.

This chapter identifies and explains the commonly used standard forms of construction contract. It considers the suitability of these forms for some of the various development situations. In conclusion, it offers a strategy for effective contract selection.

Standard forms of construction contract

The nature of standard forms of construction contract

Standard forms of construction contract have been developed to provide formal, predetermined arrangements and mechanisms to cope with the situations that can arise during the course of a construction project. They define the obligations and liabilities of the parties to the contract. They also allocate risks to the parties, different standard forms of contract generally allocating risks differently between contractor and Employer. Business is often described as the process of taking risk in return for reward. Contractors will accordingly price the risks that a contract requires them to bear, and this price will be included within their tenders. It is not therefore, as sometimes thought, necessarily advantageous to transfer as much risk as possible to the contractor. Decisions should be made, as part of the procurement route and contract selection process, about which party is most suited to manage each of the individual risks. The chosen contract strategy for any project should reflect these decisions. Murdoch and Hughes[1] note that 'any decision about laying off risks on to others must involve weighing up the frequency of occurrence against the level of premium being paid for the transfer'. The choice of procurement and contract strategy should be that offering most value to a client, bearing in mind their objectives.

The 'family hierarchy' of contracts

The various standard forms of construction contract can be considered in terms of a 'family hierarchy'. The body responsible for drafting individual standard forms can be considered as the first level in the hierarchy of standard forms of construction contracts. So, for example, the contract examined in detail in this book, JCT Standard Building Contract with Quantities (SBC/Q), has been written by the Joint Contracts Tribunal for the Standard Form of Building Contract. JCT SBC/Q is one form of contract which sits within the family of contracts produced by the Joint Contracts Tribunal. This family can be considered as the subsequent level within the hierarchy of standard forms of contract. Moving down the hierarchy from main forms of contract lie the various sub-contract forms, supplements, warranties and other ancillary documents.

Contract selection is, therefore, concerned with, first, selecting the appropriate family of contracts and, secondly, identifying the suitable main form and ancillary documents from within that family. The available families of standard contracts include the following:

1 *The Joint Contracts Tribunal (JCT)* – JCT contracts are considered by many as the 'industry standard'. They are certainly among the most comprehensive, and they are the most widely used standard forms of contract. The JCT family of contracts covers most forms of procurement and building types, and there is an impressive collection of ancillary documentation published to support the main forms. The JCT comprises interest groups from all sectors of the industry; its objective is to determine the format, content and wording of standard forms of contract. As such, JCT contracts are considered by many to be fair in that they are not loaded in favour of either party to a contract. They are, however, sometimes described as compromise conditions that, in trying to satisfy the interests of all, are unnecessarily long and complex. The variety of forms available from the JCT will be considered in more detail later in this chapter.
2 *Association of Consultant Architects' Form of Building Agreement* – The ACA form of contract was first published in 1982 by the Association of Consultant Architects. The third edition was published in 1998 and consequently revised in 2003. The ACA have also produced a form of sub-contract to complement the main contract. The forms originated in response to the Association's dissatisfaction with JCT contracts, in particular, with the allocation of risks and responsibilities. Unlike JCT contracts, this form of contract has not been created by negotiation. The form is flexible in use; by appropriate selection of alternative clauses it can be used for both traditional procurement (with or without quantities) and design and build or develop and construct procurement methods. However, the

form is not widely accepted, and its use has been minimal. The ACA also produce a partnering contract PPC2000, amended 2003.

3 *GC/Works Contracts* – The General Conditions of Government Contracts is a form of contract often used for central government works. It is not a negotiated form of contract. The forms are published by HMSO who provide a suite of contracts to cover a number of procurement arrangements, building works and services, some examples are provided below:
   - GC/Works/1: Part 1 Major Works with quantities.
   - GC/Works/1: Part 2 Major Works without quantities.
   - GC/Works/1: Part 3 Major Works single stage design and build.
   - GC/Works/1: Part 6 Construction Management.
   - GC/Works/2 Minor Works for use on small or simple contracts.
   - GC/Works/3 M&E works.
   - GC/Works/7 Measured Term Contract
   - GC/Works/10 Facilities Management.

4 *The British Property Federation System* – The BPF System was introduced in 1983. The system presented radical changes to the way buildings are procured. The system adopts an amended version of the ACA form of contract. The system has received very little use.

5 *CIOB Contracts* – Prior to its amalgamation with the CIOB in 2003, the Architects' and Surveyors' Institute (ASI) produced the ASI Building Contract. It suffered from the usual concerns of non-negotiated forms of contract in that some of the conditions may be considered as unfair contract terms. A standard form of sub-contract was published for use with the main form. Chappell[2] notes about the ASI Building Contract that 'it is intended for use on large projects, but one must doubt the wisdom of doing so in view of the ambiguous wording and inadequacy of some of the provisions'. Following the amalgamation, the CIOB produced a number of contracts in 2004, CIOB Building Contract for use with substantial and major works, CIOB Domestic Sub-contract and CIOB Minor Works Contract.

6 *ICE Contracts* – The ICE Contract is published by the Institution of Civil Engineers and has been prepared by the Institution of Civil Engineers in conjunction with the Association of Consulting Engineers and the Federation of

Civil Engineering Contractors. It is a negotiated form of contract recommended for use on major civil engineering contracts. It can be used for public or private contracts. The sixth edition, ICE6, was first published in 1991 and is a re-measurement contract; contractors are paid at contract rates on the basis of work undertaken. There is a standard form of bond to accompany the contract. The ICE has also produce a seventh edition (Measurement version) as well as contracts for minor works and design and construct.

7 *FIDIC Conditions of Contract* – The Conditions, produced by the International Federation of Consulting Engineers, are intended for use internationally. They are broadly based upon the ICE Conditions. In 1998 FIDIC updated its series of colour-coded contracts:
   - FIDIC Conditions of Contract for Construction (red).
   - FIDIC Conditions of Contract for Plant and Design and Build (gold).
   - FIDIC Conditions of Contract for EPC Turnkey Projects (silver).
   - FIDIC Short Form of Contract (green).

8 *The Engineering and Construction Contract* – The first edition of the contract (called the New Engineering Contract) was published by the Institution of Civil Engineers in March 1993. Publication followed a review of alternative contract strategies for civil engineering work which was aimed at identifying good practice. Following publication of Sir Michael Latham's report,[3] in which adoption of the form was recommended, amendments were made to the contract and a second edition published in November 1995 with the new title – The Engineering and Construction Contract (ECC), and a third edition was subsequently published in 2005.

   The objectives in drafting the form were threefold. First, to create a contract that was flexible in use. The ECC is comprehensive; it is designed to cater for all types of project, forms of procurement, methods of tender, and for use in any country. Second, to offer a contract which is clear and simple in terms of language and structure. Third, to provide a contract which acts as a stimulus for good management. The ECC system of construction contracts is designed to operate on a different basis from most of the UK contract systems. The principle adopted is that the Engineering and Construction Contract provides nine core clauses which apply to any procurement route chosen by the client. The core clauses cover issues such as the contractor's main responsibilities, time, testing and defects, insurance and disputes. The client is then given the choice of six main options that determine the procurement route to be used, i.e.:

- Option A Priced contract with activity schedule.
- Option B Priced contract with bill of quantities.
- Option C Target contract with activity schedule.
- Option D Target contract with bill of quantities.
- Option E Cost reimbursable contract.
- Option F Management contract.

The client then has a choice of up to twenty optional clauses that may be incorporated into the agreement with the contractor, these include issues such as: bonds and guarantees, sectional completion, design liability, fluctuations and damages. This 'pick and mix' approach allows the client to tailor the contract documentation to suit his particular needs.

Results from RICS surveys[4] seem to indicate that the use of the ECC has not been widespread although anecdotal evidence seems to indicate that a number of large client organisations are making extensive use of this contract system. The contract has, however, received good reviews from many within the industry, especially concerning the way it challenges many of the contemporary problems prominent within the construction industry. The contract offers a challenge to the JCT family of contracts, and it will be interesting to monitor how it affects their use.

The JCT family of contracts

The Joint Contracts Tribunal produces a wide range of standard forms of contract for use in different situations. They are most suitable for building, as opposed to engineering, projects. Surveys by the RICS[5] have shown that JCT contracts are by far the most widely used standard forms of building contract in the UK. The main forms, together with some of the associated ancillary documents, are considered below.

1 *JCT Standard Building Contract* – There are three basic variants of this JCT building contract: 'with quantities', 'without quantities' and 'with approximate quantities'. All variants

require Employers to appoint designers to carry out the design function.

The 'with quantities' edition (SBC/Q) is intended for use where the work is designed prior to contract and a bill of quantities has been prepared setting out the quality and quantity of the works. Contract documents comprise the form of contract, contract drawings, and contract bills. This is a lump sum form of contract; there is an agreed contract sum to be paid to the contractor by the Employer. Contractor's risk is limited to price only – the Employer bears the risk of errors in the bill of quantities.

The 'with approximate quantities' standard form of contract (SBC/AQ) enables construction to commence prior to the design being completed. In practice, a significant proportion of the design decisions must be made prior to contract. Approximate bills of quantities set out the quality and approximate quantity of work. This is a re-measurement form of agreement – the actual work required is measured and priced in accordance with the mechanisms and prices included in the agreement. There is no contract sum. The contract value is ascertained after the contract has been agreed.

The 'without quantities' standard form of contract (SBC/WQ) is intended for use where the design is completed prior to contract but where there is no bill of quantities. The contract documents will normally include drawings, specification and a schedule of rates. The costs to contractors of tendering for 'without quantities' contracts are higher than for contracts based upon bills of quantities because contractors have to analyse the design and ascertain the quantities required themselves. For this reason, it is unusual, except perhaps in the case of negotiated contracts, to use this form for high-value projects. Contractor's risk includes both price and quantity. This is a lump sum form of contract.

Although these contracts can be used on projects of any size, it is unusual, bearing in mind the alternative forms available from the JCT, to use these forms for projects with a value below £200,000. The contracts are comprehensive in providing for many eventualities, but suffer from being complex in wording and numbering. The contracts are subject to regular amendment and updating by the JCT to take into account developments in the law and practice.

Prior to 2005, the JCT published 'Private' and 'Local Authority' versions of the above contracts, but this approach has now been amended and both classes of employer are now covered within a single version of the contract. Optional clauses are used within the contract to accommodate the differences between the two basic employer organisations. Also prior to 2005, the JCT published two supplements that could be used to modify the standard forms of contract to allow for sectional completion or contractor designed work. Since 2005 these separate supplements have been integrated into the contract conditions in the following manner:

(i) *Sectional Completion* – for use where a contract is to be completed in phases. The contract recital identifies whether there is to be provision for the work to be completed in sections. The contract particulars, contract drawings and contract bills will then provide the necessary detail to enable this option to operate effectively.

(ii) *Contractor's Designed Portion* – for use where the contractor is required to undertake part of the design. If an employer wishes the contractor to be responsible for designing and completing a portion of the works, then the relevant information must be inserted into the seventh recital of the contract. Otherwise the seventh to tenth recitals should be deleted.

Since 2001, the JCT has produced a standard form of sub-contract which main contractors employed on one of the above contracts may use when appointing domestic sub-contractors. Its use is not mandatory and contractors may use alternative forms of sub-contract agreements if they wish. The documentation comprises two forms, the Agreement (SBCSub/A) and the Conditions (SBCSub/C). If the sub-contractor is required to carry out some design work then an alternative Agreement (SBCSub/D/A) and Conditions (SBCSub/D/C) should be used.

2 *JCT Minor Works Building Contract (MW)* – MW 05 is designed for use on projects which are short in duration, small and simple. The contract is short and easy to follow. The same form can be used for both private employers and local authorities. MW 05 is a lump sum form of contract; design should be completed by the Employer's representative (usually an architect or building surveyor) prior to a contract being agreed. There is no provision in the form for the use of bills of quantities; contract documents will usually include a combination of drawings, specifications

and schedules of works. Practice Note M2 suggested that the form was suitable for contracts with a value of up to £70,000 (at 1987 price levels); this is a guide only, and the Minor Works form has been used successfully on contracts with a value far greater than the recommended limit. The disadvantages of the form – or perhaps the reason why it is only intended for small, simple projects – lie in its lack of comprehensiveness. The provisions dealing with claims are limited, and there are no provisions for fluctuations. If the client wants the contractor to carry out some of the design work then the Minor Works Building Contract with contractor's design (MWD) should be used.

3 *JCT Intermediate Building Contract (IC)* – The Intermediate form was introduced by the JCT to fill the gap between the complex JCT Standard Building contract and the simpler Minor Works contract. The form is suitable for projects that are simple in content, require only basic trades and skills, where the services installations are not complex, there are no specialist installations, and the works have been designed by the Employer prior to tender. As a guide, it is recommended that the contract is used where the contract period does not exceed one year, and the contract value is below £280,000 (at 1987 price levels). IC 05 can be used for both private employers and local authorities and it includes an option to allow sectional completion. It is flexible in terms of the documentation that may be used in conjunction with it; together with drawings, the form permits the use of specifications, or schedules of work, or bills of quantities. The form permits sub-contractors to be named by the Employer. The following supporting documentation is produced for use with IC 05:
   (i) *Named Sub-contract documentation* – ICSub/NAM is the form of tender and agreement and ICSub/NAM/C are the conditions of sub-contract for use when there are named sub-contractors. ICSub/NAM/E is an agreement for use between the Employer and a sub-contractor and its use is recommended where a named sub-contractor has a design responsibility.
   (ii) *Domestic sub-contract documentation* – ICSub/A is the form of agreement and ICSub/C are the sub-contract conditions. If the domestic sub-contractor is required to execute some design work as part of the package, then alternative forms, Agreement ICSub/D/A and Conditions ICSub/D/C should be used.

4 *JCT Design and Build Contract (DB)* – DB 05 is intended to be used when the contractor is responsible for design and construction. It is similar in content and complexity to the JCT Standard Building Contract. The contract is flexible in that it caters for both private and local authority employers and that it permits design input by the Employer up to tender stage. The JCT has also produced a set of standard forms of sub-contract that may be used with this procurement route: Design and Build Sub-Contract Agreement (DBSub/A) and Design and Build Sub-Contract Conditions (DBSub/C). There is no mention of an architect or quantity surveyor in the contract; contract administration is performed by a duty holder referred to as the Employer's agent (in practice, the role of Employer's agent could be undertaken by a surveyor, an architect, an engineer, a project manager or a combination of these disciplines).

In essence, the contract operates such that tenders are invited on the basis of satisfying a set of Employer's Requirements. The nature of the Employer's Requirements can, in practice, vary from brief performance requirements to detailed designs. Tenders are submitted in the form of Contractor's Proposals (the contractor's response to the Employer's Requirements). These subsequently form the basis for the contract.

The form received immediate, widespread use on its introduction, and its popularity has continued to grow. The form provides employers with the reassurance of single point responsibility for design and construction. It provides a low risk arrangement for employers so long as they comply with the spirit of the form and keep post-contract variations to the Contractor's Proposals to a minimum. The reasons for popularity may also include effective marketing of the design and build process by design and build contractors; and dissatisfaction from employers with the unpredictability of performance using other procurement and contractual arrangements.

5 *JCT Management Building Contract (MC)* – Under MC 05, a Management Contractor is appointed to arrange and co-ordinate construction work. The Management Contractor does not carry out any of the work; this is undertaken by a series of Works Contractors. The Employer is required to arrange for the design to be carried out by a design team. The Management Contractor acts as an interface between the designers and the Works Contractors. He is paid on the basis of a fee for managing the

Works, together with the actual cost (or prime cost) of the Works Contracts plus the cost of providing site facilities. The Employer's financial commitment will not be known before work commences, and probably not until a project is nearing or has reached completion – effective cost management and forecasting is therefore fundamental in achieving an Employer's financial objectives. Employers bear much of the financial risk under this form of arrangement. The contract does, however, offer the opportunity of commencing work on site prior to the design being completed. This should mean an overall reduction of the development programme for a project when compared with traditional procurement approaches. Other advantages include:

(i) the ability to let Works Contracts on the basis of competition, which should ensure that the overall project cost is representative of market prices;
(ii) in a falling market, Works Contracts will be representative of the falling market (they are normally progressively let as the programme requires) and thus the Employer should receive any associated financial benefits;
(iii) in a rising market, Works Contracts will be representative of the rising market, and perhaps the likelihood of contractor insolvency, and the detrimental effect of insolvency on the success of a project, will be reduced;
(iv) contractor input can only add to the efficiency of designs;
(v) ability to alter the project requirements during construction.

The benefits offered by management contracts are greatest where projects are high in value, long in duration and complex. The JCT produces the following documentation for use in conjunction with MC 05:

- *Management Works Contract Tender and Agreement (MCWK)* – comprising Contract Tender and Agreement.
- *Management Works Contract Conditions (MCWK/C)* – the conditions of contract between the Management Contractor and a Works Contractor.
- *Management Works Contractor/Employer Agreement (MCWK/E)* – an agreement between the Employer and a Works Contractor.

6 *JCT Construction Management* – where the client wishes to appoint a construction manager the appropriate forms are: the

Construction Management Tender (CM) and the Construction Management Agreement (CM/A). As explained in the previous chapter, one of the main features that differentiates construction management from management contracting is the fact that the client contracts directly with the trade contractors and for this purpose the JCT have produced the Construction Management Trade Contract (CM/TC). Also included within this suite of documents are two collateral warranties: CMWa/F for use between a trade contractor and a company providing the client with project finance, and CMWa/P&T for use between a trade contractor and a tenant or purchaser of the building works.

7 *JCT Prime Cost Building Contract (PCC)* – The Prime Cost Contract was originally introduced to replace the Fixed Fee Form of Prime Cost Contract. It is broadly based upon the provisions of the JCT Standard Building Contract. The contractor is paid the cost of carrying out the work plus a fee to cover profit and overheads. The Employer is responsible for arranging the design. Although the design does not have to be completed prior to work starting on site, the design must be developed sufficiently to enable the contractor to ascertain the appropriate fee. The fee can be either fixed or paid on the basis of a percentage of the prime cost. Where the fee is fixed, the contract provides for adjustment of the fee if the difference between the actual prime cost and the estimated prime cost is greater than 10%. The Employer bears the majority of financial risk under this form of contract. The total cost of a project will not be known until completion.

PCC 05 should be used in situations where it is not possible to define the extent of the works prior to tender, either due to lack of time or because the nature of the work is such that it cannot be defined (for example, some types of repair and refurbishment work).

8 *JCT Measured Term Contract (MTC)* – The measured term contract is intended to be used where an Employer requires maintenance or minor works to be undertaken on a regular basis. Contracts are usually agreed on the basis of a schedule of rates for carrying out certain types of work on a defined list of properties. The rates are used to value the work carried out over a defined period of time. The employer engages a contract administrator and quantity surveyor to oversee the running of the contract. Contracts can be let on a fixed or fluctuating

price basis. The standard form contains a break provision for terminating the contract early.

9 *JCT Major Project Construction Contract (MP)* – The first edition of this contract was published in 2003 with the advice that it was designed for use on projects that were significant both in size and complexity. Surprisingly the contract itself is shorter and simpler in detail than other corresponding JCT contracts. The reason being that the contract is designed for client organizations who have their own in-house contractual procedures which has allowed the JCT draftsmen to reduce the amount of detail within their contract conditions. The contract is flexible regarding design responsibility, e.g., a contractor may only be required to complete a detailed design provided by the employer or alternatively could be asked to provide a full design for the works along with associated risks and responsibilities. Consequently, further advice is provided that contractors employed under this contract should be experienced and knowledgeable and be able to put into place proper systems of risk management.

An approach to effective contract selection

The governing principles

The requirements and objectives of the Employer will be the primary factors driving contract selection. For most projects, the selection of procurement route to meet the Employer's needs will limit the available choices of standard form to be considered. In most situations, procurement and contract selection will be considered simultaneously.

The decision about which form of contract to use for a project will be based upon the following factors:

1 *The nature of the client* – Answers to the following questions may be pertinent to contract selection:
   (i) Are clients private individuals or organizations or public bodies?
   (ii) Are clients regular/frequent developers or is a one-off project envisaged?
   (iii) How knowledgeable are clients about their needs?

(iv) How knowledgeable are clients about the processes of construction and development?

(v) What past experience of construction have clients had, and how does their experience prejudice the use of the various alternative procurement and contract choices?

2 *The risk attitude of the client* – What risks are individual clients prepared to take in relation to the associated rewards available? This will depend to some extent on the nature of the client. Public clients may tend to be more averse to risk than private clients. The nature of development may also affect risk attitude – clients engaging in speculative commercial developments may be prepared to take greater risks with regard to the construction cost of a project in order to receive the benefits associated with achieving earlier completion.

3 *The procurement method adopted* – Some of the standard forms of construction contract are written for use in connection with specific procurement strategies while others (for example, ECC) are flexible. The source of design, whether generated by the Employer or contractor, will also affect the choice of contract.

4 *The client's priorities in terms of time, cost and quality* – The balance of requirements will affect the choice of contract.

If clients want cost certainty prior to agreeing a contract, then a lump sum form of contract is appropriate. This approach requires Employers either to allow sufficient time for their consultants to complete the design prior to a contract being agreed or to adopt a design and build approach. The first option will generally increase the overall development time, whereas the design and build approach may mean a reduction in quality.

If time is of the essence, then either a management contract, a prime cost contract or an approximate quantities contract may be appropriate, depending on the other decision factors. None of these forms of contract offer cost certainty.

Where quality is the main priority, the use of a design and build contract will probably be inappropriate.

5 *The size of the project* – The various types of JCT contract suitable for different sizes of project have been discussed above.

6 *The type of documentation being used* – Although contract selection will normally drive the requirement for contract documentation.

7 *The type of project* – Answers to the following questions will affect the decision on form of contract:
   (i) Is the project engineering or building based?
   (ii) How complex is the project?
   (iii) Does the project involve specialist installations or construction techniques?

In advising on contract strategy, it is important for surveyors to remain impartial. In many situations, the benefits – for example, higher fees – may be greater under one contract strategy than another. This must not influence the advice given. The recommended contract should be that which offers the greatest value to a client. Conversely, it may be necessary in some situations to advise clients that, although they have been offered a service by, for example, a design and build contractor that purports to offer the best solution to their needs, this may not be the case. Of course, it is in the design and build contractor's interest to sell his service.

It is often the case that a bad experience in using a particular form of contract can prejudice both a client and a consultant against using or recommending that form again. Problems and difficulties can occur under any form of contract, and it is probable that the problem is not down to the contract but is caused by some other factor. Bad experience should not necessarily preclude the use of a standard form of contract on future projects.

A strategy for selection

The wide range of available standard contract options makes the use of decision-making techniques advisable to ensure that rational selections are made. The experienced practitioner will often develop an intuitive skill in selecting an appropriate form of contract. However, for the inexperienced, a more formal approach to selection is often helpful. Indeed, where a client requires a detailed account of the reasons behind a recommended contract strategy, it is sometimes helpful to adopt a formal approach to explaining a particular rationale.

The criteria for decision-making will generally flow from the requirements of the Employer, and to ensure effective decision-making it is essential to define these criteria. There will be a difference between the wants and needs of an Employer. While it is not unreasonable for an Employer to want the best of all worlds in terms of low cost, low risk, short programme, high quality, cost certainty, etc., it is not usually possible for procurement and contract strategy to provide this. It is therefore necessary to identify Employer needs. This can be done by prioritizing those wants that are essential to the Employer. Having established this hierarchy of needs, it should be possible, in conjunction with the other decision-making criteria, such as the size and complexity of the project, to identify the contract strategy that is most suitable for each given situation. An approach similar to that described in Chapter 2 in connection with procurement is appropriate.

The selection of a contract form will normally involve identifying the most appropriate family of contracts, selecting the individual form of contract to be used, and identifying the appropriate support documentation required. This chapter has briefly described the nature of the different families of contracts together with the main forms within the JCT family of contracts. This should help in directing the surveyor to an appropriate form for a given situation. However, it will be essential for the surveyor to analyse in detail the provisions within that form prior to making a recommendation. The aspects of the form which do not match the requirements of the project should be identified. If these are minor points, then the form of contract may still be suitable, perhaps with some minor amendment or addition (but note that any changes to standard forms of contract should be made by legal experts and preferably be limited to minor changes – see comment on p. 2. If the issues are major, an alternative contract strategy should be considered. Despite the wide choice of standard contracts available, it is unlikely that any contract will exactly match all the requirements of a project; therefore, sensible compromise will often be necessary.

In giving advice on contract selection, surveyors should clarify the reasons for their recommendations, together with the conditions that must be observed in consequence of the selection. Too many projects suffer as a result of the requirements pertaining to the use of a particular contract not being met. For example, if a project is being undertaken on the basis of a JCT Standard Building Contract without Quantities, the contract envisages that the design should be practically complete prior to the contract being agreed. However, in practice it is not uncommon to witness a lump sum contract being awarded when much of the design is incomplete. This can create risk for Employers, and is unnecessary in view of the numerous alternative methods of procurement and forms of contract that are designed to be used in situations where the work is to start before the design is complete.

References

1. MURDOCH, J. and HUGHES, W., *Construction Contracts Law and Management* (London: E & FN Spon 1992), p. 15.
2. CHAPPELL, D., *Which Form of Building Contract?* (London: Architecture Design and Technology Press, 1991), p. 62.
3. LATHAM, Sir Michael, *Constructing the Team* (London: HMSO, 1994).
4. DAVIS, LANGDON and EVEREST, *Contracts in Use: A Survey of Building Contracts in Use During 2001* (London: RICS, 2003).
5. op. cit.

Bibliography

1. CHAPPELL, D., *Which Form of Building Contract?* (London: Architecture Design and Technology Press, 1991).
2. JCT, *Practice Note 20: Deciding on the Appropriate Form of JCT Main Contract* (London: RIBA Publications Ltd, 1993).
3. JCT, *Practice Note 7: Standard Form of Building Contract for Use with Bills of Approximate Quantities* (London: RIBA Publications Ltd, 1987).
4. ASHWORTH, A., *Contractual Procedures in the Construction Industry*, 2nd edition (Essex: Longman Scientific and Technical, 1991).
5. MURDOCH, J. and HUGHES, W., *Construction Contracts Law and Management* (London: E & FN Spon, 1992).
6. PIKE, A., *Practical Building Forms and Agreements* (London: E & FN Spon, 1993).

4

Tendering methods and procedures

After a client has selected an appropriate procurement strategy for his building project, the next stage will be a review of how best to obtain the resources that will be necessary for him to have the work carried out. In most instances, client organizations will have limited skills and resources relating to construction work and they will have to rely on others to provide the necessary services, expertize and resources, e.g., designers, engineers, surveyors and contractors. The client will therefore need to identify and contact suitable personnel or organizations to assist him with his project, and agree with them the scope and nature of their work or the resources to be supplied and the basis of payment. This will be achieved through a process of tendering. Tendering is an important stage of the building project, as the decisions taken at this point will help to determine the quality and calibre of people and organizations involved in the building process and the price to be paid for their services.

Over a period of time a number of publications have been issued to help clients with the tendering process. For nearly 35 years the National Joint Consultative Committee for Building (NJCC) published a number of highly regarded codes of procedure relating to tendering practices.[1] However, in the late 1990s the NJCC was disbanded and its advisory role was assumed by the Construction Industry Board (CIB) who produced its own tendering guidelines.[2] Despite the existence of these newer guidelines, many client organizations still adhere to the original NJCC codes of procedure and, when the Joint Contracts Tribunal (JCT) produced its own practice note on tendering,[3] it acknowledged the inclusion of substantial material originally produced by the NJCC.

It is as well to be aware that the tendering process is not used exclusively to appoint a contractor to carry out the client's project. It is also highly likely that most members of the project team will be asked to go through a tendering procedure. With most procurement routes, the client will want to appoint an architect and quantity surveyor to help him manage the various processes, and an appropriate tendering procedure should ensure that the client obtains a competent service at a realistic price. Once an architect and surveyor are appointed, they will then be able to assist the client in the selection and appointment of the main contractor.

Methods of contractor selection

The following recommendations are based predominantly upon the guidelines published by the JCT and the original guidelines produced by the NJCC.

The selection of a contractor to carry out a construction project is an important matter requiring careful thought. A wrong choice may lead to an unhappy client/contractor relationship, a dissatisfied client and possibly even an insolvent contractor.

Any one contractor will not be suitable for any one job. A contracting organization will be geared to work for a particular size or price range and will be unsuitable or uneconomic for contracts outside that range. For example, a national company would probably not be suitable to carry out a small factory extension or to build a single house on an infill site in an urban location. The overhead costs of that size of organization would be such as to make it impossible for the company to carry out the work for an economic price. A medium-sized company may be generally competent in most fields of building works, but may not have sufficient design expertize to take on projects that contain a substantial amount of contractor-designed work. Finally, at the other lower end of the scale, a small firm, used only to jobbing work and small alterations and extensions, would probably be out of depth in handling a contract for the erection of even a modest-sized reinforced concrete office block. Such a firm, with little experience of estimating and probably none in the pricing of bills of quantities, would be likely to tender at too low a level by failing to include for all the costs involved. If given the contract, the firm would no doubt soon get into difficulties and probably end up in disaster.

The client's professional advisers should aim to find a contracting company

1 that is financially stable and has a good business record;
2 for which the size of the project is neither too small nor too large;
3 that has a reputation for good-quality workmanship and efficient organization; and
4 that has a good record of industrial relations.

There are three principal methods of choosing a contractor: (a) open tendering, (b) selective tendering and (c) nomination.

Open tendering

This is initiated by the client's project manager, architect or quantity surveyor advertising in local newspapers and/or the technical press, inviting contractors to apply for tender documents and to tender in competition for carrying out the work, the main characteristics of which are given. Usually, a deposit is required in order to discourage frivolous applications, the deposit being returnable on the submission of a *bona fide* tender.

Advantages

1 There can be no charge of favouritism as might be brought where a selected list is drawn up (this is of concern particularly to local authorities who, probably for this reason more than any other, tend to use open tendering more than other clients).
2 An opportunity is provided for a capable firm to submit a tender, which might not be included on a selected list.
3 It should secure maximum benefit from competition (it may not always do so, however, as may be seen from the disadvantages below).

Disadvantages

1 There is a danger that the lowest tender may be submitted by a firm inexperienced in preparing tenders (particularly if bills of quantities are used), and whose tender is only lowest as a consequence of having made the most or the largest errors.
2 There is no guarantee that the lowest tenderer is sufficiently capable or financially stable. Although obtaining references

will provide some safeguard, there may be little time in which to do so.

3 Total cost of tendering is increased as all the tenderers will have to recoup their costs eventually through those tenders which are successful. The result can only be an increase in the general level of construction costs.

Open tendering was deprecated in the Banwell Report.[4] Consequently central government departments and most local authorities do not use this method of obtaining tenders. There tends, however, to be an increase in its use by local authorities in times of national economic difficulties with attendant public expenditure cuts, presumably because it is assumed that the more tenderers there are, the lower the price for which the work can be carried out.

Selective tendering

Selective tendering may be either *single* or *two stage*, depending on whether the full benefits of competition are desired (in which case single-stage tendering is used) or whether limited competition plus earlier commencement of the works on site is considered advantageous (using two-stage tendering).

Single-stage selective tendering

Under this procedure, a short list is drawn up of contractors who are considered to be suitable to carry out the proposed project. The names may be selected from an approved list or 'panel' maintained by the client (as many public authorities do), or may be specially chosen.

In the latter case the contractors may be invited, through suitably worded advertisements in the press, to apply to be considered for inclusion in the tender list. This gives the client the opportunity to exclude any firms thought to be unsuitable, and to limit the number of tenderers. At the same time, it gives any firm the opportunity to apply to be considered. The criteria that a client may use to decide upon the suitability of a contractor are:

1 Whether the company has had recent experience of similar projects of a similar standard and completed within the envisaged time scale.

2 Whether the company has the skills necessary for the delivery of the project.
3 The quality of the company management structure and personnel employed by the company.
4 Whether the company has the spare capacity to carry out the work within the proposed time scale.
5 The financial standing and record of the company.

It is recommended that the number of tenderers should be limited to between five and eight, depending on the size and nature of the contract. If the firms on the list are all ones which are reputable, well-established and suitable for the proposed work, and the client fixes the construction time, then the selection is resolved into a question of price alone and the contract can be safely awarded to the firm submitting the most favourable tender.

Advantages

1 It ensures that only capable and approved firms submit tenders.
2 It tends to reduce the aggregate cost of tendering.

Disadvantage

1 The cost level of the tenders received will be higher, owing to there being less competition and also due to the higher calibre of the tenderers.

Nomination

This is sometimes referred to as *single tendering* and is, in effect, a special case of selective tendering, the short list containing only one name. It is used when the client has a preference for a particular firm, often because it has done satisfactory work for him before.

Obviously, if only one firm tenders for a job, competition is eliminated and that will, almost inevitably, lead to a higher price. The client may think it is worth paying more, however, in return for a quicker job or one of better quality than he might otherwise get. One should be fairly certain, however, that a worthwhile benefit will accrue before advising a client to nominate rather than go out to competitive tender.

When a contractor is nominated, the contract sum will be arrived at by a process of negotiation. This may be done using bills of quantities or schedules of rates, but instead of the contractor pricing the tender document on his own and submitting his tender to be accepted or rejected, the rates and prices are discussed and agreed until eventually a total price is arrived at which is acceptable to both sides.

Usually the negotiation will be conducted between one of the contractor's senior estimators and the surveyor (either a partner, associate or senior assistant). To facilitate the procedure, one party will usually price the tender document first of all, to provide a basis for the negotiation. The other party will then go through the rates and prices, ticking off those which are acceptable and then the two surveyors will meet to negotiate the unticked ones. When agreement on the whole is reached, a contract will be entered into between the client and the contractor.

Serial tendering

Sometimes, when a large project is to be carried out in successive phases, a combination of selective tendering and nomination is employed. This is sometimes called *serial tendering*. The contractor is chosen for the first phase by means of selective competitive tendering. The accepted tender forms the basis of payment for the resulting contract in the normal way. The tender is also used for the second and later phases, provision being made for so doing in the initial contract by the inclusion of a formula for updating the prices. Alternatively, the contract for the first phase may specify negotiation of new rates, based upon the tendered prices, as the means of determining the payment for each successive phase in the series.

The purpose of serial tendering is to gain the benefits of continuity. The contractor for the first phase of the project will have his site organization set up, his offices, mess and storage huts, etc., already in use and plant of various kinds on the site. When the second phase commences, these facilities will be already available, thus allowing a smooth transition with much less additional expense than if a different contractor were to be employed.

In addition, the contractor's workforce will be familiar with the details of the construction after building the first phase, and thus should be able to work more speedily and efficiently on the second and subsequent stages.

Two-stage selective tendering

This procedure is used when it is desired to obtain the benefits of competition and at the same time to have the advantage of bringing a contractor into the planning of the project, thus making use of his practical knowledge and expertise. It may also result in an earlier start on site.

The first stage aims to select a suitable contractor by means of limited competition. The second stage is a process of negotiation with the selected contractor on the basis of the first-stage tender.

First, a short list of tenderers is prepared, as described for single-stage tendering. The NJCC Code[5] recommended a maximum of six names on the list (four in the case of specialist engineering contracts) and also suggests matters for consideration when drawing up the list.

First stage

When being invited to tender, tenderers are informed of the second-stage intentions, including any special requirements of the client and the nature and extent of the contractor's participation during the second stage. Tenderers are asked to tender on the basis of any or all of the following:

(i) an *ad hoc* schedule of rates, consisting of the main or significant items only;
(ii) a detailed build-up of prices for the main Preliminaries items;
(iii) a construction programme showing estimated times and labour and plant resources which would be used, and also construction methods;
(iv) details of all-in labour rates and main materials prices and discounts which would go into the build-up of the detailed tender;
(v) percentage additions for profit and overheads;
(vi) proposed sub-letting of work, with additions for profit and attendance.

During this stage, discussion with each of the tenderers may be conducted in order to elucidate their proposals and to enable the contractors to make any suggestions with regard to design and/or construction methods. When these procedures have been concluded, a contractor is selected to go forward to the second stage. It is important that in accepting the first-stage tender, the parties define procedures for either of them to withdraw, should the second-stage negotiations prove abortive, and what, if any, payment may be due to either party in that event, including reimbursement of the contractor for any site works the contractor may have carried out.

Second stage

During this stage, finalization of the design proceeds in consultation with the selected contractor, and bills of quantities (or other detailed document describing the proposed works) are prepared and priced on the basis of the first-stage tender. Negotiation on the prices will follow until agreement is reached and a total contract sum arrived at, when the parties will enter into a contract for the construction works.

When time is pressing and it is desired to start work on site before final agreement is reached, a contract may be signed earlier. In that case, when the surveyor considers that a sufficient measure of agreement has been reached on the prices for the principal parts of the work, he will recommend the client to proceed with arrangements for the signing of a contract. This is not recommended in the NJCC Code, but the client may consider that it is worth taking a risk in order to speed up the project.

Tender documents

The number and nature of the tender documents used will vary depending upon the procurement route and the type of contract chosen. They will include some or all of the following.

Conditions of the contract

This document sets out the obligations and rights of the parties, and the detailed conditions under which a subsequent contract will operate. If a standard form, such as the JCT Form, is used it will not be sent out with the invitation to tender, it being assumed that the tenderers will have a copy or can readily obtain one. The clause headings will, however, be listed in the first (Preliminaries) section of the bills of quantities and/or specification.

Bills of quantities

Since the 1990s, the use of bills of quantities has seen a steady decline and these days they are most likely to be used for lump sum contracts of more than say, £100,000.[6] Tenderers should be sent two copies of the bills,[7] one for return to the architect or surveyor with the tender, the other for the contractor to keep as a copy of his submitted prices. If the tenderers are not required to submit priced bills with their tenders, only one copy need to be sent to them initially. A second copy will then be sent at the time of requesting submission of the priced bills.

Specification

In the case of a traditional procurement route, utilizing a lump sum contracts without bills of quantities, a detailed specification will be supplied to tenderers. Sometimes a specification will be supplied in addition to bills of quantities where they are used, but this is exceptional nowadays. A specification may be in a form for detailed pricing.

Schedule of works

As an alternative to a specification in the case of 'without quantities' contracts, tenderers may be supplied with a Schedule of Works. This lists the work comprised in the contract under appropriate headings. The tenderers may be required to price the schedule.

Drawings

Normally, general arrangement drawings will be provided, showing site location, position of the building(s) on and means of access to the site and floor plans and elevations. Tenderers are not normally given working drawings as they are not considered to be necessary for pricing purposes, full descriptions of the work being incorporated in the bills or specification. Tenderers are informed, however, that they can inspect drawings not supplied to them, at the architect's office.

Employer's Requirements

Where a client wishes the contractor to have a design input on the project, perhaps by using a design and build procurement route or using an alternative route which contains a contractor's design input, it will be necessary for the client to clearly set out his requirements to enable the contractor to produce a suitable design. The JCT refer to this document as the Employer's Requirements.[8] There is no standard form of employer's requirements but the JCT Practice Note provides advice as to the content and detail that should be included. The purpose of the form is to provide tendering contractors with a clear idea of what the client wants in the way of a building, e.g., type of structure, function, size, accommodation, quality, aesthetics, costs in use requirements, etc. The actual detail provided in the form can vary considerably depending upon how the client wishes to use the design and build process. The form may contain only basic information as to the required function of the building, thereby allowing tenderers a free rein regarding their design proposals. Alternatively a client may have very clear ideas about his requirements, in which case the form may contain a full scheme design prepared by the client's architect, and the tenderers would be left with the task of developing the working drawings and being responsible for delivery of the scheme design. Some of the items identified by the JCT for possible inclusion in the Employer's Requirements are:

1 The current state of planning permission and who is responsible for obtaining permissions where consent has not yet been received.
2 A statement as to the function of the building(s).
3 A statement of site requirements, e.g., site boundaries, use of site, ground conditions and availability of services.
4 The level of design, structural and specification detail to be provided by the tenderers.

Contractor's proposals

The contractor's proposals are the contractor's response to the employer's requirements and will be the key document for the client to consider at the tender review. The advice provided by the JCT on the basic content of the proposals is that they should contain the following:

1 Plans, elevations, sections and typical details.
2 Information about the structural design.
3 Layout drawings, incorporating details of services to be provided.
4 Specifications for materials and workmanship.

Form of tender

This is a pre-printed formal statement in which a tenderer fills in the blank spaces typically providing his name, address and the sum of money for which he offers to carry out the work shown on the drawings and described in the bills of quantities or specification. The JCT provides a Model Form of Tender for main contract works in Appendix C to its *Practice Note 6*.[9]

Return envelope

Each tenderer should be supplied with a pre-addressed envelope clearly marked 'Tender for (name of project)'. This will ensure that tenders are recognized as such when received and will not be prematurely opened.

Tenderers should be asked to acknowledge in writing receipt of the tender documents.

Tendering procedure

Preliminary enquiry

It is a recommended practice (although not always followed) to send, about a month beforehand, to each of the firms from whom it is proposed to invite a tender, a preliminary enquiry[10] to ascertain that they are willing to submit a *bona fide* tender. This avoids the situation of contractors declining to tender or, if they prefer not to decline although not wanting to tender, submitting a 'cover price', i.e., a price which is high enough to be well above the lowest tender.

Sufficient information about the project should be given in the preliminary enquiry letter to enable each contractor to decide whether he is in a position to comply and, if so, whether he would be able in view of other commitments to carry out the contract if it is awarded to him. The JCT provides a model form of preliminary enquiry and a project information schedule to be filled in on the client's behalf. Completion of this schedule ensures that the contractor is made aware of the contract conditions to be used for the project, which optional clauses are to be used and whether there are any amendments to the standard form. The contractor is also informed whether the contract is to be entered into as a simple contract or as a deed and whether there are any requirements for the contractor to provide collateral warranties.

If for any reason a firm which has signified its willingness to tender is not included in the final short list of tenderers, it should be informed of the fact immediately, as other tender invitations may be under consideration.

Period for preparation of tenders

It is important that tenderers are given sufficient time to make all necessary enquiries from suppliers, sub-traders, etc and the date for return of tenders should be fixed so as to allow for the amount of work likely to be involved in preparing a tender for a job of the size and character of the proposed project. Four to eight weeks is recommended as a minimum period, although it is possible, in exceptional cases, that a lesser time may be adequate. Where a tender is being submitted on a design and build basis, a contractor will have to carry out considerably more work in the preparation of his tender submission and this fact should be reflected in the time allowed to return the tender, i.e., a minimum of 12 weeks.[11]

A time of day (often noon) should be specified as the latest time for tenders to be received on the date fixed. Any tenders arriving later should be returned and should be excluded from the competition.

Parity of tendering

It is of paramount importance that all tenders should be based on the same information. Consequently, all the tender documents must be identical. It would be possible, for example, for some copies of a drawing to contain the latest amendments, while other copies are of the unamended drawing. Not infrequently, a tenderer will telephone the architect or surveyor about an item in the bills of quantities, employer's requirements or a clause in the specification. He may question the accuracy of a quantity or have been told by a supplier that a specified material is no longer marketed. Such queries must, of course, be answered and must be dealt with promptly. But, whatever the reply given to the enquirer, if it adds to or varies in any way the information given in the tender documents, it must be communicated to all tenderers. This should be done immediately by telephone and then confirmed in writing. The same procedure must be followed if an error is discovered by the surveyor in the bills or specification or other tender documents or if it is decided to extend the time for the receipt of tenders. Tenderers should be asked to confirm in writing the receipt of every written communication of additional or varied information immediately.

Equally in pursuance of the principle of parity of tendering, bills of quantities should be priced as drawn up by the surveyor and likewise where tenders are based on specifications or schedules of rates. A tenderer sometimes wants to price for an alternative material which he considers to be just as good as that specified but which he can obtain more cheaply. Or he may want to allow for an alternative form, method or order of construction. The procedure, which he should adopt in such a case is to price the tender documents as printed and to submit, in an accompanying letter or other suitable form, details of the alternative material or form of construction etc. with the consequential effect on his tender sum.

If a tender is qualified in any respect which it is considered has given the tenderer an unfair advantage over the others, or which makes the comparison of the tenders unreasonable, then that tenderer should be given the opportunity of withdrawing the qualification without amending the tender sum.[12] If he is unwilling to do so, his tender should be rejected. This approach ensures that the competitive tendering process is fair to all contractors and should be considered a cornerstone of good practice. It is desirable that the number of variables in tenders should be reduced as far as possible so that they are made more readily comparable. One of these variables is the length of the construction period, and it is common practice for this to be specified by the client.

If a contractor wants to offer to do the work in a shorter time, then the procedure which he should follow is the same as for any other qualification to the printed documents, as described above. If the contractor objects to the period as unreasonably short, however, he should raise the matter with the architect during the tender period. If it is decided to extend the construction period, then all tenderers must be informed, as this may well affect their tender sums. If at the outset the client is prepared to accept alternative tenders, e.g., allowing the contractors to specify an alternative completion date, the JCT recommend that a separate form of tender be completed for each alternative.

It is obviously far more difficult to reduce the number of variables when dealing with design and build tenders, as contractors will be bidding not only on price but also design, aesthetics, function, structure, costs in use, etc. In these instances the client needs to have a clear idea of the criteria to be used to determine which is the best value bid.

Opening tenders

Tenders are normally returnable to the architect or project manager (sometimes to the quantity surveyor). A formal procedure should be followed for opening them to eliminate any suspicion of irregularities. No tender must be opened before the latest time for submission, and the specially-marked envelopes supplied with the tender documents are intended to eliminate accidental opening. There can be no possibility then of anyone communicating to another tenderer the amount of a competitor's tender.

As little time as reasonably possible should be allowed to elapse before opening tenders and they should all be opened at the same time, preferably in the presence of the architect and quantity surveyor and, if he so wishes, the client. In the case of public bodies, tenders are usually opened by, or in the presence of, the chairman of the committee responsible for the project.

Notifying tenderers

A tender list should be drawn up and sent to all tenderers as soon as possible after the meeting at which the tenders were opened. This is so that each contractor may know whether his tender was successful or not, and so be better able to judge his future commitments and know how to respond to any other invitation to tender. The list should contain all the tenderers' names arranged in alphabetical order and all the tender sums in ascending order. It should not disclose which tenderer submitted which amount but, of course, each will be able to identify the position of his own tender in relation to the lowest. At the same time, all but the three lowest should be informed that their tenders have not been successful. The reason for this approach is to safeguard the client's position if for any reason the original successful tenderer withdraws his offer, e.g., the lowest tenderer may have made a large error in the pricing document and which forces him to withdraw his tender. The second and third tenderers should be informed as soon as the tender has been accepted. The preparation of the tender list and its communication to tenderers is a duty, which often falls to the surveyor to carry out.

Contractor selection; quality versus price?

Usually in a competitive tendering situation, price tends to be the only criterion for selection however where a contractor has more involvement with the project through a design or management input it may be advisable to consider the quality of the tender submission as well as the price to determine which tender provides the best value. If such an approach is to be adopted then the tendering organizations should be fully informed of the selection criteria and the relevant weighting to be applied to each criterion. The JCT in its project information schedule identifies possible additional criteria that may be considered in a contractor's tender submission, i.e.

1 Approach, e.g., method statement, programme.
2 Customer care, e.g., liaison with employer.
3 Environmental, e.g., proposals for reduction in noise and nuisance.
4 Mangement, e.g., health and safety.
5 Resources.
6 Sub-contractors and supply chain, e.g., calibre and length of time of business relationships.
7 Technical.

8 Design and build, e.g., aesthetics, life cycle costs, flexibility in use.

Liability associated with tendering

Most private and commercial organizations have little legal liability when asking companies to submit a tender; the general principle is that an organization requesting a tender is not obliged to accept the lowest tender or indeed any tender submitted. This fact is often reinforced in the invitation to tender letter sent out by clients, e.g., 'The employer reserves the right to ... accept any tender or no tender at all. No tendering expenses will be payable.'[13] Although tenderers may incur considerable costs in preparing and submitting tenders, this has to be viewed as their commercial risk with the possible reward that work may be obtained if the tender is successful. Therefore as long as there is no negligent misrepresentation in the tender documentation (*J Jarvis and Sons Ltd v Castle Wharf Developments Ltd (2001) 4 Ll Rep 308*) and the request for tenders is genuine, i.e., the person asking for tenders at the time of enquiry was genuinely considering letting the work, the tenderers will have no claim against the client if no tenderer is appointed or the project is subsequently shelved.

However, organizations such as public and local authorities may be treated slightly differently when it comes to the tendering process. In the case of *Blackpool and Fylde Aero Club Ltd v Blackpool B. C. (1990) CA 1 WLR 1195* the local authority failed to open and consider a tender that had been delivered on time. By an oversight of the council, the tender form was not collected from its post box and the tender was subsequently declared to be late and void. The decision of the Court of Appeal was that when a party correctly submitted a competitive tender for public works, the authority receiving the tender had a contractual duty to consider it along with all other compliant tenders. Therefore there had been a breach of that duty and the authority was liable for damages.

Tendering and the EU

Public and local authorities also need to be aware of the Public Works Contract Regulations which were introduced to some extent as a result of directives from the European Union. The Regulations are intended to ensure that, when public works are put out to tender, the criterion for the award of the contract is to be clearly stated. This is to ensure that there is fair and open competition amongst member states. For example, the award of a contract may be based on the lowest price submitted or alternatively the most economically advantageous tender (EU Directive 93/37, Article 30). However, where the latter is to apply, the contracting authority must state all the criteria it intends to apply when awarding the contract. Failure to comply with this Directive may lead to a complaint from an unsuccessful tenderer as in the case *of Harmon CFEM Facades (UK) Ltd v The Corporate Officer of the House of Commons (2000) 72 ConLR 21.* In this case it was held that Harmon had not been treated fairly, that there had been a breach of the Public Works Contract Regulations and consequently Harmon were entitled to damages and costs. Eventually the House of Commons reached an out of court settlement with Harmon. As on December 2000, the cost to the House for failing to follow the correct tendering procedures was £9,896,429, which comprised a combination of agreed damages and legal costs.[14]

Electronic tendering

In 1998, the JCT produced a guidance manual for parties who may be interested in the use of electronic data interchange (EDI) within the construction industry.[15] Also, since 2000 the Government has been keen to promote the use of e-tendering in the procurement of goods, services and capital projects, both in the public and private sectors. Therefore, it is apparent that the start of the twenty-first century is likely to see a growth in the use of electronic communications and data interchange within the construction industry.

When a client wishes to receive tenders through the EDI process, the JCT recommends that all the tenderers should be consulted beforehand to obtain their agreement to this procedure. If a tenderer does not agree to the tender being conducted through EDI, they should not be removed from the tender list but should be provided with hard copies of the tender documents. The JCT approach to EDI is to try and encourage firms to adopt this procedure and not to coerce them into complying, therefore it was considered unreasonable to exclude a company from the tender list merely because of its refusal to accept electronic tendering. Surprisingly the JCT recommends that despite conducting the tender process through EDI, the actual forms of tender should be submitted as hard copy. The reason being that they were of the opinion that at that time current technology was not sufficiently sophisticated to prevent unauthorized access to financially sensitive information which may in turn lead to corrupt practices.[16]

Procedures

A basic requirement of electronic tendering is for all the appropriate forms, documents and drawings to be converted into an electronic format which may then be uploaded onto a server. Tenderers will then be provided with the necessary information to allow them to access the server and download the documentation. Some consideration needs to be given when deciding upon the format of the various items of electronic data, i.e., it needs to be in a format that can be easily read by the tenderers' software. Safeguards also need to be built into the system to prevent the unauthorized amendment or alterations of the electronic data. Obviously there will be occasions when the client's team may need to alter or amend certain tender documentation, in which case there needs to be a secure system of access, control and permissions to ensure that only authorized personnel are allowed to carry out this task. To complement this safeguard there also needs to be a clear system of identifying updated documentation and of notifying all interested parties of the changes. If the tenderers are required to email their form of tender to the client, it is recommended that they are automatically stored on a secure area of the server that will not allow the emails to be accessed until the due time and then only by authorized personnel. This process should help to overcome some of the concerns expressed earlier by the JCT regarding the security of electronically transmitted forms of tender. The tenderers may then be promptly informed of the outcome of the tender opening meeting by email.

Some of the claimed advantages of using electronic tendering are:

1 It may be possible to reduce the normal tender period, as documents, alterations notifications can be transmitted in a matter of seconds as compared to days when using postal or courier services.

2 There should be a reduction in the client's and tenderers' copying and mailing costs. Although computers have not yet created the paper free office, and without careful control a surprising amount of paper can still be produced.
3 It should be possible to quickly respond to email queries, which in turn can be swiftly filed and archived which should aid the efficient administration of a project.
4 It may be possible to speed up the estimating process. In one case study it was demonstrated that a contractor tendering on a £25 m project was able to save six days in his estimating department because all the documentation was available in an electronic format that was imported into his estimating system in a matter of minutes.[17]
5 The estimating process may also be made more efficient by the fact the contractor having accessed and imported the electronic data into his estimating package may then quickly break it down into work packages which in turn can be swiftly transmitted to his sub-contractors for them to price and return electronically.
6 On receiving the contractor's tender documents the client's team may then use appropriate software to swiftly analyse the bid, prepare cost analyses and cost forecasts. The price data may also be subsequently used to assist in the preparation and checking of interim valuations and pricing variations.
7 General administration should be improved for all parties as all the documentation should be stored electronically, allowing authorized access at any time and from any place that has a communications link to the server.

On-line bidding

On-line bidding is a variation from the normal tendering process. The general procedure is for tendering organizations (who may be asked to pre-qualify) to view and download the tender requirements and specification on-line. The bidding process will commence at a stipulated time with the client setting an opening price for the tenderers to bid against. The principle being that the tenderers will compete by underbidding, i.e., submitting lower prices, which is why the system is sometimes referred to as an 'electronic reverse auction'. There are various ways by which the tenderers can follow the progress of the auction depending upon how it has been set up by the client. A basic requirement is that the tenderers can view the bidding process in real time, as a minimum they should be able to see the lowest bid submitted (where price is the only criterion) or the most advantageous bid (where price is not the sole criterion). Further information may be displayed such as the number of firms bidding in the auction and the ranking of a company's bid. Tenderers may submit as many bids as they like, as frequently as they like, up until the specified closing time of the auction. It is normally recommended that tenderers be informed that if a new bid is submitted within 30 minutes before the set closing time of the auction then the closing time the will be automatically extended, by a defined period, to allow other tenderers to respond if they wish. This procedure should help eliminate the ploy of 'auction sniping' which often occurs on other on-line auction sites such as E-Bay. As soon as the set time has been reached and there have been no last-minute bids the auction is closed. If the tender is to be awarded on price alone then the lowest tenderer will have immediate feedback of his success.

The use of electronic reverse auctions for the procurement of capital projects or professional services is rather controversial. Supporters of the system claim it an efficient and transparent means of awarding contracts. Opponents of the system claim it is nothing more than an electronic version of a 'Dutch auction', a process that is certainly employed in the construction industry but usually considered to be contrary to good practice. Opponents also claim the system is contrary to the principles of trust, partnering and best value, ideals which the construction industry is trying to encourage in an endeavour to improve output and quality of the industry.

Where an electronic reverse auction is used and cost is the sole criterion, experience tends to show that the quality of the service or product supplied may be adversely affected. A tenderer who has submitted a rock bottom quotation may subsequently try and improve its negligible profit margins. This may be by attempting to supply goods or services of a lesser quality, or by combing the tender documentation to find loopholes or errors that would allow extras to be claimed. Therefore it is important for the client's team to ensure that the tender documentation is clear and unambiguous, the specification is clearly defined and standards and outcomes can be given with a high degree of precision.

References

1. NJCC, *Code of Procedure for Single Stage Selective Tendering* (London: RIBA Publications, 1996).
2. CIB, *Code of Practice for the Selection of Main Contractors* (London: Thomas Telford, 1997); CIB, *Code of Practice for the Selection of Subcontractors* (London: Thomas Telford, 1997).
3. JCT, Series 2, *Practice Note 6 – Main Contract Tendering* (London: RIBA Publications, 2002).
4. *The Placing and Management of Contracts for Building and Civil Engineering Work* (London: HMSO, 1964).
5. NJCC, *Code of Procedure for Two Stage Selective Tendering* (London: RIBA Publications, 1996).
6. Davis, Langdon and Everest, *Contracts in Use* (RICS, 2003).
7. NJCC, *Code of Procedure for Single Stage Selective Tendering*, op. cit.
8. JCT, *Practice Note CD | 1A* (London: RIBA Publications, 1981).
9. JCT, Series 2, *Practice Note 6 – Main Contract Tendering*, op. cit.
10. CIB, *Code of Practice for the Selection of Main Contractors*, op. cit.
11. Ibid.
12. JCT, Series 2, *Practice Note 6 – Main Contract Tendering*, op. cit., p. 8.
13. Ibid., p. 20.
14. *Appropriation Accounts 1999–2000*, New Parliamentary Building, Portcullis House: Losses (London: National Audit Office, February 2001).
15. Electronic Data Interchange in the Construction Industry, Joint Contracts Tribunal (London: RIBA Publications, 1998).
16. Ibid., p. 91.
17. Get started in e-Business, Construct IT, www.construct-it.salford.ac.uk.

5

Examining and reporting on tenders

Examination of selected tender

The following advice is worded on the basis that the client's criterion for selection is price, where this is not the case and the client has used additional criteria then references to lowest tender should be read as best value tender.

Once the lowest tender has been identified, the surveyor will proceed to look at it in detail. There will usually be supporting documents, such as schedules of rates, contract sum analysis and, in many cases, bills of quantities requiring careful scrutiny. If there are bills, the lowest tenderer will be asked to submit his priced bills for examination (if not sent with his tender), at the same time being informed that his was the lowest tender and it is therefore under consideration.

To save time in the event of the lowest tender proving unsatisfactory, the next lowest, or the next two lowest tenderers, may also be asked to submit their priced bills. They should be informed that although their tenders were not the lowest, they are under consideration. Opinions vary among surveyors as to whether the second or third lowest tenderer's bills should be looked at until the lowest has been rejected. Some say they should not, as being of no concern at all should the lowest tender be accepted. Others take the view that there may be errors in the second or third bills, which, if corrected, would make one of them the lowest. From the basic legal point of view that it is the originally tendered total sums which are all-important, it seems logical that, even if the three (or two) lowest tenderers' bills are called for, the second and third ones should not be looked at until the lowest tender has been rejected. Should the lowest be satisfactory, the other bills should be returned to the contractors unexamined.

The surveyor, when examining bills of quantities, schedules of rates, etc., will look for errors of any kind and for any anomalies or peculiarities, which in his opinion, make it unwise for the client to enter into a contractual relationship with the tenderer. It should be remembered, however, that while it is made clear to tenderers when tendering that the client does not bind himself to accept the lowest or any tender, contractors naturally expect that there will be sound reasons for rejecting the lowest tender. The surveyor must therefore be able to provide those reasons in the event of the lowest tender not being accepted.

It has already been stated that the important factor in a lump sum tender is the total sum for which the contractor offers to carry out the work. If that offer is accepted, then, legally, a binding contract exists. It may be asked therefore 'Why examine the tender in detail'? It is because of the uses made of the component parts of a tender subsequently that any errors in the bills of quantities/schedules of rates need to be identified and satisfactorily dealt with. Otherwise, problems will almost certainly arise during the course of the contract and in the settlement of the final account. For example, an error may have occurred in a bill when multiplying the quantity of an item by the unit rate and thus an incorrect amount for that item is shown. If the error was not corrected and subsequently the item was affected by a variation order, say, halving the quantity, what is to be done? Should half of the total amount be omitted or should the surveyor omit half of the quantity at the unit rate shown? The latter course could result in the omission of an amount, which is greater than the incorrect total for the item. It is to avoid such difficulties that errors should be found and dealt with in a satisfactory manner.

A further reason is that the examination of the tender details is likely to reveal whether an arithmetical error of such a magnitude has occurred that it would not be wise to hold the contractor to the tender sum because he would then be working at a much reduced profit (assuming the tender sum to be too low). If such an error were in the contractor's favour, then it would be unfair to the client to have to pay more than the job was worth.

The things, which the surveyor will look for when examining bills or schedules of rates will be of the following kinds.

Arithmetical errors

These may occur in the item extensions (i.e., the multiplication of the quantities by the unit rates), in the totalling of a page, in the transfer of page totals to collections or summaries, in the calculation of percentage additions on the General Summary or by rounding-off the bill total when carrying it to the Form of Tender.

Pricing errors

These are patent errors *not* matters of opinion as to whether a rate is high or low. A patent error will sometimes occur where, in the transition from cube to superficial measurement, or from superficial measurement to linear, the estimator has continued to use the basic cube rate for the superficial items, or the basic superficial rate for the linear items. Such errors are obvious and not the subject of opinion. Another type of patent error is where, apparently by an oversight, an item has not been priced at all which one would normally expect to be priced.

Again, identical items in different sections of the bills may have been priced differently without there being any good reason. These differences usually arise because the estimator has not remembered that an identical item has occurred before or has not bothered to look back to see what rate had been used. This type of error is not uncommon in housing contracts, where there are separate bills for each house type, and in elemental bills where items recur in different elements.

Pricing method

Sometimes a tenderer will not price most, or any, of the 'Preliminaries' items, their value being included in some or all of measured rates. This might create difficulty should any adjustment of the value of 'Preliminaries' become necessary.

Some contractors may price all the measured items at net rates (i.e., exclusive of any profit and overheads), showing the latter as an inserted lump sum on the General Summary page of the bills. This is unsatisfactory when using the bill rates later on and the tenderer should be asked to distribute the lump sum throughout the bills or to agree to its being treated as a single percentage addition to all the prices.

Again, a contractor may price the groundwork, *in situ* concrete and masonry sections of the job (i.e., those sections which will largely be completed in the early months of the job) at inflated prices, balancing these with low rates in the finishing sections. His object would be to secure higher payments in the early valuations, thus improving his cash flow position. This procedure would tend to work against the client's interest, particularly if the contractor were to go into liquidation before the contract was completed. Such a pricing policy might possibly indicate a less stable financial situation than previously thought.

The contractor is entitled, of course, to build up his tender as he thinks fit and cannot be required to change any of his rates or prices. However, contractors usually recognize and accept the need to correct obvious errors and the surveyor would be imprudent not to do so. As regard to pricing strategy, this is usually considered to be a matter for the contractor's decision, the remedy being, if the surveyor has genuine grounds for concern about it, for him to recommend the client not to accept the tender.

Further points to note in regard to tenders, which are not based on bills of firm quantities, are as follows:

Tenders based on bills of approximate quantities

The rates for items of which the quantities are likely to be substantially increased upon measurement of the work as executed should be carefully scrutinized, as these may affect the total cost of the job considerably.

Tenders based on drawings and specification only

The schedule of the principal rates which have been used in the preparation of the tender should be looked at closely as these rates will be used in the pricing of any variations which may be subsequently ordered. If there is likely to be much dayworks, the percentage additions to prime costs, which the tenderer has quoted, may also become significant.

Tenders based on contract sum analysis

On a design and build project the contractor will be asked to provide a contract sum analysis. The analysis will provide only a basic breakdown of the contractor's lump sum. The detail of the breakdown can be very limited with the contractor apportioning his tender to the various elements of the building, e.g., design work, preliminaries, groundworks, concrete, masonry, cladding, etc. In this instance it is not possible to carry out a detailed analysis but the figures should be reviewed to ensure that the allocation appears to be realistic.

Tenders based on ad hoc schedule of rates

The rates for items, which are likely to be significant in terms of quantity, need careful scrutiny. The above comment in regard to dayworks will also apply here.

Reporting on tenders

The surveyor must report to the project manager or the architect and the client as soon as the examination of the tenders is complete. He must remember that the purpose of the report is to enable the client to decide whether to accept any of the tenders and, if so, which one. Consequently, it should concentrate on matters of importance and minor matters, such as the details of arithmetical errors, should be excluded.

The form of the report will vary according to the nature of the tender documents but usually will include the following:

(a) the opinion of the surveyor as to the price level, i.e., that the tender is high, low or about the level expected;
(b) the quality of the pricing, indicating any detectable pricing method or policy;
(c) the extent of errors and inconsistencies in pricing and the action taken in regard to them;
(d) the details of any qualifications to the tender;
(e) the likely total cost of the project, if not a lump sum contract; and
(f) a recommendation as to acceptance or otherwise.

The examination and reporting on tenders for *design and build contracts* (see p. 34) is not as straightforward a matter as for other types of contract. There will be a number of variables, making comparison between tenders difficult. Apart from price differences, they will vary in the construction time, constructional form, quality of finishes, degree of maintenance likely to be required subsequently and

KEWESS & PARTNERS
52 High Street, Urbiston, Middlesex, UN2 1QS.

Shops and Flats, Thames Street, Skinton, Middlesex

Report on tenders

1. Tender list. Six tenders were received as follows:

Contractor	Tender Sum
Beecon Ltd	£1,448,000
W.E. Bildit Ltd	£1,505,980
E.A. Myson Ltd	£1,521,192
Thorn (Contractors) Ltd	£1,540,207
Smith & Sons Ltd	£1,574,433
Brown & Green Ltd	£1,685,013

   The lowest tender is 3.85% below the second lowest tender. It is also within the amount of our last amended cost plan dated 20 May, 2005, which was in the sum of £1,504,000.

2. Bills of Quantities.

   The priced bills of quantities submitted by the lowest tenderer, Beecon Ltd., have been examined and the following is a report on our findings.

3. Pricing.

   3.1 The general level of pricing is, in our opinion, slightly lower than average for this class of work, even allowing for the present economic climate. We do not consider, however, that this should be a cause for concern in this instance as Beecon Ltd. are well-established contractors in the area with a reputation for good quality work. Also, it is known that they particularly wanted to obtain this contract because of its prominent location and its proximity to their head office.

   3.2 The pricing is consistent throughout with no serious anomalies or detectable adverse characteristics.

   3.3 A number of arithmetical errors were discovered and have been corrected in accordance with the provisions set out in the tender document. Beecon Ltd. have elected to stand by their tender in the sum of £1,488,000 notwithstanding the errors. Had the tender sum been amended, it would have been reduced by £29,161, which is the net amount of the errors.

4. Recommendation. We recommend that the contract be awarded to Beecon Ltd. in consideration of the sum of £1,488,000.

5 August, 2005.

Figure 5.1 Report on tenders

perhaps the anticipated cost of running and maintaining specialist installations. In his report, the surveyor will need to consider all the tenders since there will not be a lowest tender as such. He will need to set out in a clear and easily assimilated form the relative savings, benefits, merits and demerits between the packages offered. He should, nevertheless, make his recommendation as to which one, if any, the client should accept.

Figure 5.1 is a report on the tenders for the contract described in Appendix A and illustrates the application of some of the foregoing principles.

Dealing with errors, etc.

In the case of tenders not based on bills of quantities, the scope for detection of errors, etc. is limited and correction of any which are discovered is usually a straightforward matter of adjusting any erroneous rates, with the contractor's agreement.

If a lump sum tender based on drawings and specification only shows a breakdown of the total and this contains arithmetical errors, then correction should be done in a similar way to that described for bills of quantities, so far as is applicable.

Two alternative ways of dealing with errors in bills are recommended by the Joint Contracts Tribunal (JCT).[1] The first is the traditional procedure of informing the tenderer of the errors and asking him whether he wishes to confirm his tender, notwithstanding the errors (which may be favourable or unfavourable to him), or whether he wishes to withdraw it. This procedure accords with the basic principle of the law of contract whereby an offer is made to provide goods and/or services for a certain consideration and the offerer may withdraw his offer at any time before its acceptance. The recommendation of the JCT is that this method of dealing with errors is inappropriate for two stage-selective tendering and possibly partnering arrangements, and in these instances the client may be advised to consider adopting the second alternative.

The second course of action is to inform the tenderer of the errors and to ask him whether he wishes to confirm his tender, notwithstanding the errors, or whether he wishes to correct genuine errors. If he chooses to correct his errors and the revised tender sum is then higher than the next on the list, then his tender will be set aside and the lowest one will be proceeded with in a similar way instead.

It should be noted that the term 'genuine errors' means inadvertent arithmetical errors and patent pricing errors, as defined earlier. It does not include errors, which the estimator may have made in the build-up of a bill rate or price. The provision for the correction of genuine errors does not give the contractor the opportunity of increasing any rates or prices that on reflection he realizes are too low, nor of pricing any items which were intentionally left unpriced.

One reason why the second alternative has been introduced is that, because of the large sums involved in most construction projects, it is in clients' interests to allow the correction of such errors as have the effect of increasing the tender sum, if to do so leaves the tender still in the lowest position. The contractor will be happier and the client will be paying a lower price than if the contractor felt he had no alternative but to withdraw. In any case, it is not in the client's interest to have a discontented contractor from the start of the contract, who feels that he is beginning the contract with an expectation of a reduced profit. However, it must also be acknowledged that the adoption of the second alternative could allow a contractor to manipulate the situation to his advantage.

The JCT recommends that it should be decided from the outset which of the alternative methods of dealing with any errors will be used, and this decision will be communicated in the preliminary enquiry to tenderers, the formal invitation to tender and the Form of Tender.

When the contractor elects to confirm his tender, his errors must still be dealt with so that the priced document is arithmetically correct and consistent throughout. This is so that when the rates and prices are used subsequently, difficulties that would otherwise arise will be eliminated (see p. 90, under the heading '*Examination of selected tender*'). The correction of errors must, of course, be done in consultation and agreement with the contractor. The surveyor will normally draw up a list of errors, etc., which he will agree with the contractor, and the surveyor will then amend the priced document accordingly.

The procedure for the correction of errors in bills is as follows:

(a) If the original text has been altered without authorization, the alteration(s) should be deleted and prices adjusted accordingly.

(b) All errors of the kinds described earlier (i.e., arithmetical and patent pricing errors) should be corrected and the corrections carried through from page totals to collections, from collections to summaries and from summaries to general summary. The incorrect amounts should be ruled through with a single line and the correct ones written above, neatly and clearly.

(c) A sum, equal to the net amount of the aggregated errors, should be inserted on the General Summary immediately before the final total as an addition or deduction, as appropriate. This will have the effect of adjusting the corrected bill total to the amount stated on the Form of Tender (which, it should be noted, in some instances may not be the same sum as the original total of the bills).

This net addition or deduction equal to the errors is, in effect, the aggregate of adjustments to every rate and price in the bills, excluding any contingency sum, provisional sums and Preliminaries (if deducted). Accordingly, any subsequent use of a rate or price from the bills or of one based upon rates or prices in the bills must be subjected to an equivalent adjustment. To facilitate this, the balancing sum (i.e., the net amount of the errors) is calculated as a percentage of the value of the builder's work. The contract copies of the priced bills should be endorsed with a statement that all the rates and prices (other than contingency and provisional sums) are to be considered as increased or reduced (as appropriate) by that percentage. The example in Figure 5.2 illustrates how the process may be affected.

All surveyors do not agree on what constitutes 'builder's work' for the purpose of calculating the adjusting percentage referred to in the last paragraph. The JCT says that in addition to the fixed sums mentioned above, the value of Preliminaries items should also be excluded. Turner[2] does not agree, taking the view that the latter are just as much the contractor's own prices as those applied to measured items. This seems a reasonable view, particularly as arithmetical errors may occur just as easily in the Preliminaries as in the measured sections of the bills.

In the calculation on page 98, the Preliminaries have been deducted from the contract sum in accordance with the JCT advice. The bills were those submitted in support of the tender for the contract described in Appendix A, in which the corrected amounts are shown in the Summary of Bill No 3 and in the General Summary.

List of errors in priced bills submitted by Beecon Ltd

Page no.	Item ref.	Type of error	Effect of error on total of tender: Increased by £	Decreased by £
25	–	Page total	300.00	
30	–	Page total		18.00
48	C	Item extension	195.60	
79	–	Page total transferred incorrectly		1.35
86	L	Item extension		56.50
103	D	Item extension	460.00	
140	–	Page total		1,500.00
General Summary		Page total transferred incorrectly from p. 141	54.00	
General Summary		Sub-total	30,000.00	
Form of Tender		Total rounded down		272.75
			31,009.60	1,848.60
			1,848.60	
		Net amount of errors	29,161.00	

The following pages show how the error in item D on p. 103 of the bills was dealt with and the consequential correction of totals, also the corrections to the Summary and General Summary of all the errors listed above. On the General Summary is shown the sum added to balance the net error and also the endorsement. The adjusting percentage referred to in the endorsement has been calculated thus:

$$\frac{29{,}161}{1{,}418{,}839.00 - (126{,}507.00 + 50{,}688.00 + 24{,}297.17)} \times 100 = 2.395\%$$

This calculation should be shown on the General Summary page in a convenient position or on the facing page.

Figure 5.2 Example of correction of errors in bills of quantities

L40 GENERAL GLAZING

	STANDARD PLAIN GLASS				£ *p*
	OQ quality sheet glass				
A	4 mm Panes, area 0.15–4.00 m² to wood with bradded wood beads.	71	m²	17.85	1,977 35
B	4 mm Ditto but bedded in glazing compound.	60	m²	25.98	1,558 80
C	4 mm Ditto, area not exceeding 0.15 m² (In 110 nr. panes).	16	m²	29.01	464 16
D	4 mm Ditto, area 0.15–4.00 m² to metal with mastic and spring glazing clips.	46	m²	25.98	1,195 08 ~~1,655 08~~
	Rolled glass, pattern group 2				
E	3 mm Panes, area 0.15–4.00 m², to wood with bradded wood beads.	20	m²	25.13	502 60
	Rolled Georgian wired rough cast glass				
F	6 mm Panes, area 0.15–4.00 m², to wood with bradded wood beads.	48	m²	42.25	2,028 00
	Rolled Georgian wired polished plate glass				
G	6 mm Panes, area 0.15–4.00 m², to wood with bradded wood beads	17	m²	54.53	927 01
H	6 mm Ditto, area not exceeding 0.15 m², ditto (In 140 panes).	20	m²	54.86	1,097 20
	GG quality clear plate float glass				
J	6 mm Panes in sliding hatch doors, 475 mm wide × 570 mm high, ground on all edges,1 nr. ground and shaped finger grip sinking.	20	nr	28.57	571 40
					10,321 60
				Carried to Collection	~~£10,781 60~~

103

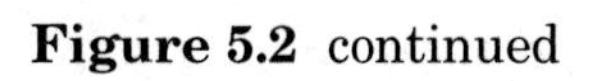

Figure 5.2 continued

L40 GENERAL GLAZING (CONT)

	STRIPS FOR EDGES OF PANES				£ p
A	Glazing strips as described to edges of 3 mm glass.	188	m	1.08	203 04
B	Ditto but 4 mm glass.	320	m	1.08	345 60
C	Ditto but 6 mm glass.	406	m	1.08	438 48
		Carried to Collection		£	987 12

COLLECTION		£ p
Page	93	2,560 20
"	94	825 74
"	95	1,317 09
"	96	2,150 65
"	97	5,792 20
"	98	1,746 51
"	99	3,107 26
"	100	2,329 15
"	101	2,834 21
"	102	2,244 77
"	103	10,321 60 ~~10,781 60~~
"	104	987 12

WINDOWS/DOORS/STAIRS Total Carried to Summary £ 36,216 50 ~~36,676 50~~

104

Figure 5.2 continued

Bill No. 3 – Shops and flats

Summary

Page		£
26	Groundworks	181,837.93 ~~182,137.93~~
33	In situ concrete	184,537.67 ~~184,519.67~~
44	Masonry	155,400.25
69	Structural/carcassing metal/timber	69,930.48 ~~70,126.08~~
79	Cladding/covering	32,051.34 ~~32,049.99~~
92	Waterproofing	64,103.00 ~~64,046.50~~
104	Windows/doors/stairs	125,291.63 ~~125,751.63~~
109	Surface finishes	52,447.88
112	Furniture/equipment	13,598.22
116	Building fabric sundries	32,051.36
121	Disposal systems	29,137.62
124	Piped supply systems	21,045.45
129	Mechanical heating	84,214.65
134	Communications/security/control	22,325.56
136	Electrical services	32,051.79
140	Transport systems	71,235.00
	Total carried to general summary £	1,171,259.83 ~~1,172,139.58~~

141

Figure 5.2 continued

General summary

Page		£
15	Bill No. 1 Preliminaries/general conditions	126,507.00
19	Bill no. 2 Demolition	5,531.00
141	Bill no. 3 Shops and flats	1,171,259.83 ~~1,172,193.58~~
147	Bill no. 4 External works	40,556.00 ~~39,056.00~~
151	Bill no. 5 Provisional sums	50,688.00
		1,394,541.83 ~~1,423,975.58~~
Add for:		
	Insurance against injury to persons and property as item 5C	10,000.00
	All risks insurance as item 5C	12,920.17
	Water for the Works as item 13A	1,377.00
		1,418,839.00
	Add to adjust for errors	29,161.00
	Total carried to form of tender	1,448,000.00 ~~1,448,272.75~~

All rates and prices herein (excluding preliminaries and provisional sums) are to be considered increased by 2.395%

Signed .... Signed ....

152

Figure 5.2 continued

Where there is only a small mathematical error in the tender documents, for example, a few hundred pounds on a £5 m project, some surveyors adopt a pragmatic approach to the adjustment. With the agreement of the contractor, the error adjustment may be accommodated within the Preliminaries bill. The price of one or two preliminary items may adjusted by the total of the error, the Preliminaries bill would then be recalculated so that the contract sum in the bill of quantities is correct and in accordance with the contractor's original tender figure.

References

1. JCT, Series 2, *Practice Note 6 Main Contract Tendering* (London: RIBA Publications, 2002), p. 10.
2. TURNER, DENNIS F., *Quantity Surveying Practice and Administration, Third Edition* (London: George Godwin Ltd., 1983), p. 125.

6

The contract

Under the general law of contract, when a party makes an offer to provide goods and/or services for some certain consideration and the party to whom the offer is made accepts it, then, provided it does not involve any illegal act, a contract which is enforceable at law exists. This is no less the case in the construction world than in any other sphere of business or industry. The offer is made by a contractor who tenders to carry out specified construction works in return for a money payment and upon the acceptance of that offer by the client promoting the project, a binding contract comes into being.

It is desirable that the client's acceptance should be in writing and that it should be given as soon as possible after receipt of tenders. If there is any appreciable delay of, say, one to two months or more, then it will be necessary to obtain the contractor's confirmation that his price still stands or what the increase in his price is, due to rising costs.

The contract documents

The number and nature of the contract documents will normally correspond with those of the tender documents (see p. 75). They may be some or all of the following.

Form of contract

This is the principal document and will often be a printed standard form, such as one of the variants of the JCT Form. For example, JCT 05 SBC/Q comprises two parts: (i) the Agreement, which is where the parties sign and become legally bound to the contract. (ii) The Conditions (subdivided into nine sections including attached schedules), which set out the obligations and rights of the parties and detail the conditions under which the contract is to be carried out.

Bills of quantities

Any errors in the bills must be corrected in the manner described in Chapter 5 and any necessary adjustments to rates and prices must be clearly and neatly made.

Specification

Under the 'with quantities' variants of the JCT Form, the specification (if one exists) is not a contract document but it is under the 'without quantities' variants. In the latter case, the specification may have been prepared in a form for detailed pricing and have been priced by the contractor when tendering. If this is not the case, the contractor may have been asked to submit with the tender a Contract Sum Analysis. This shows a breakdown of the tender sum in sufficient detail to enable variations and provisional sum work to be valued, the price adjustment formulae to be applied and the preparation of interim certificates to be facilitated.

JCT Practice Note 23[1] gives a guide to the identification of specified parts of the contract sum and the breakdown of the remainder into parcels of work. A Contract Sum Analysis may not be required, where a Schedule of Rates is used. In cases where a Contract Sum Analysis or a Schedule of Rates is provided, it is signed by the parties and attached to the contract, but is not a contract document. However, the specification is a contract document in such cases.

Schedule of works

Where bills of quantities do not form part of the contract (for example, under the JCT Standard Form without quantities), the contractor may be sent a Schedule of Works for pricing when tendering, instead of a specification. If so, the priced Schedule of Works will become a contract document.

Schedule of rates

Where bills of quantities are not provided, a schedule of rates is usually necessary as a basis for pricing the work in measured contracts and for pricing variations in the case of lump sum contracts. Such schedules are described on pp. 21–22.

Drawings

The contract drawings are not limited to those (if any) sent to the contractor with the invitation to tender but are all those which have been used in the preparation of the bills of quantities or specification.

It is important that the contract drawings are precisely defined, as amendments may be made to them during the period between inviting tenders and commencing the work on site. Such amendments may affect the value of the contract and, if they do, should be made the subject of Architect's Instructions. It is good practice therefore, for the surveyor to certify in writing, on each of the drawings used for taking-off or specification purposes, that they were the ones so used.

Preparations for executing the contract

The surveyor is often regarded by architects as the expert in matters relating to the interpretation and application of the contract Conditions. Accordingly, it is not uncommon for the preparations for the signing of the contract to be the task of the surveyor.

Each of the contract documents is important from a legal point of view. Both parties (and their agents) are bound by what is said in them. When a contract proceeds satisfactorily to completion and settlement, many of the details in those documents may not be referred to. They become all-important, however, when a contract runs into difficulties and particular statements in the documents may then come under close scrutiny. If there are ambiguities, discrepancies or contradictions in them, it may lead to delays, adjudication, arbitration proceedings or actions in the Courts, all of which are expensive as well as frustrating.

If such faults in the contract documents are due to the carelessness of the persons responsible for their preparation, their principals (i.e., the partners in the firm of surveyors or architects, as the case may be) may be liable for a claim for damages for negligence, which might prove very costly. Great care must be exercised therefore, in carrying out the task of preparing the documents for the execution of the contract.

Each of the contract documents as already listed will require attention, if only to facilitate the actual signing, and they will now be considered in turn.

Form of contract

The Articles of agreement

For example, with the JCT 05 SBC/Q contract the blank spaces will be filled in with the date the agreement was made, the names and addresses of both the Employer and the contractor, a brief statement of the nature and location of the Works, the numbers of the contract drawings, the contract sum (in words and in figures) and the names and addresses respectively of the architect/contract administrator and quantity surveyor. Also, where the contractor is to design a portion of the works, the nature of the design works is to be specified.

On contracts where bills of quantities are not used (i.e., JCT 05 SBC/XQ), it is possible that the employer would not appoint a quantity surveyor, as a result an Article is provided which gives the option of naming the person who shall be exercising the quantity surveying functions. The option not required should be struck out in a neat and clear manner.

After the Articles of Agreement and the section on Contract Particulars, there follows the Attestation where there are spaces for the signatures of the parties. Where the agreement is to be executed under hand the Employer (or his representative) signs in the first space and the contractor (or his representative) signs in the second space and the signature of a witness follows their signature. If the signature of the Employer and contractor is obtained at the same time, as is usual, the witness may be the same person in each instance.

When either party must, or wishes to, execute the contract as a deed, as in the case of some local authorities and other public bodies and corporations, an alternative signature page is provided where there are three options by which the agreement may be executed. A corporate company may execute the contract by imprinting its common seal on the document in the presence of a director and company secretary (or two directors) or alternatively by inserting the names of a director and company secretary (or two directors). In both instances these actions are to be confirmed by the signature of the director company secretary, etc. A local authority or similar organization that is not covered by the Companies Act 1985 must only use the first option, i.e., the use of a common seal. The third option relates to a party who is not a corporate organization and is acting as an individual, in which case they are to sign in the appropriate place in the presence of a witness who likewise is to sign the contract.

The Conditions

Some clauses of the Conditions in the JCT Forms contain alternative wording. Attention is drawn to these alternatives in footnotes and to the need for appropriate deletions. For example, clauses 4.8, 4.17.4, 4.19, 4.21, 6.7 provide just some instances. Such deletions should be clearly and neatly made.

There may be need or desire for other alterations or additions to be made to the clauses of a standard form in a particular case. All such alterations and additions, as well as the alternatives to be deleted, should be specifically listed in the 'Preliminaries' section of the bills or specification and, if this has been done, the list may then be used as a checklist in the preparation of the standard form for executing the contract.

The blank spaces in the Contract Particulars (see Appendix B) should also be filled in, in accordance with a further list in the bills or specification. It should be noted that if nothing is stated against certain of the clause numbers, then specified time periods will apply.

The parties should initial each amendment to the printed document in the adjacent margin. This serves to confirm that all amendments were made before the contract was executed and that no amendment was made subsequent to that event.

Bills of quantities

The priced bills submitted by the contractor should show clearly all the amended and corrected rates and prices, etc., which have been agreed between the contractor and the surveyor. If an adjusting amount, equal to the amount of the errors, has been entered in the General Summary, then an endorsement should be added indicating the percentage effect on the measured rates and prices (see p. 102). On the reverse of the last page a further endorsement should be added, saying 'These are the Contract Bills referred to in the Contract' with spaces for the signatures of the parties, set out in the same way as in the Articles of Agreement.

Specification

Where the specification is a contract document, all that is necessary is to endorse on the reverse of the last page the words 'This is the Specification referred to in the Contract' with spaces for the signatures as in the case of bills. The schedule of rates should show clearly all corrections and amendments which have been agreed between the surveyor and the contractor. If, alternatively, a Contract Sum Analysis has been submitted, this should be signed by the parties and attached to the Specification or to the Contract Conditions.

Schedule of Works

If a priced Schedule of Works has been submitted by the contractor when tendering, that will be a contract document instead of the specification. It should be endorsed in a similar manner as that described for specifications and be signed by the parties.

Drawings

It is desirable that each of the drawings should be endorsed either 'This is a contract drawing' or 'This drawing was used in the preparation of the bills of quantities' or some similar statement. Many firms have rubber stamps for this purpose.

Following the execution of the contract, the contract documents are to be kept by the Employer although the contractor has the right to inspect them at any reasonable time.

Copies of contract documents

Immediately after the execution of the contract, the contractor should be given a certified copy of the Articles of Agreement, the

Conditions of the Contract, the contract drawings and the contract bills. In practice, the two copies of these documents are prepared at the same time so that they are identical, including all the initialling and the signatures. Where the contract is sealed, only the Employer's copy needs to bear the seals.

The contractor should be given in addition, two copies of the unpriced bills, or, if there are no bills, two copies of the specification, and two copies of the drawings, plus two copies of any descriptive schedules or similar documents. If the surveyor is responsible for organizing the arrangements for the signing, he will usually collect together beforehand two sets of the documents and hand them to the contractor at the meeting at which the contract is executed.

The contract sum

In the case of a lump sum contract, a specified sum of money is stated in the contract as to be paid by the Employer to the contractor for the carrying out of the construction work. This sum is seldom, if ever, a fixed amount but is subject to additions and deductions in respect of various matters. Consequently, the JCT Form, in the Articles of Agreement, after providing for the insertion of the contract sum, continues 'or such other sum as shall become payable under this Contract'.

The main causes for adjustment of the contract sum for which the Conditions make provision are: the adjustment of provisional sums; variations to the design and/or the specification of the work; additions or reductions to the scope of the work, loss or expense incurred by the contractor for specified reasons and increases or decreases in the costs of labour and materials or in taxes, levies or contributions imposed by Government.

The variable nature of the contract sum may, at first sight, appear to nullify the value of stating any sum at all. It does have the merit, however, of giving the parties a good indication of the level of cost and providing a basis for estimating the eventual cost at later stages in the progress of the project.

The scope for adjustment of the contract sum is not unlimited. Clause 4.2 of the JCT Form limits adjustment to the express provisions of the Conditions of the contract. It specifically excludes the correction of errors made in the contractor's computation of the contract sum.

References

1. JCT, *Practice Note 23, A Contract Sum Analysis* (London: RIBA Publications, 1987).

Bibliography

1. KEATING, D., *Keating on Building Contracts*, 6th edition (London: Sweet & Maxwell, 1995).

7

Variations and instructions

Definition and origin of variations

The definition of the term 'variation' in clause 5.1 of the JCT Form indicates the wide scope for the exercise of the architect's power to vary the Works.

Variations may arise in any of the following situations (references are to the JCT Form):

(a) when the architect needs or wishes to vary the design or the specification (see clause 5.1.1);
(b) when a design deficiency is discovered in the Employer's Requirements (see clause 2.14.2 and 2.14.3);
(c) when a discrepancy is discovered between any two or more of the contract documents (see clause 2.15);
(d) when a discrepancy is discovered between any statutory requirement and any of the contract documents (see clause 2.17.2.2);
(e) when an error in or omission from the contract bills is discovered (see clause 2.14.1 and 2.14.3);
(f) when the description of a provisional sum for defined work (see page 000 in the contract bills does not provide the information required by General Rule 10.3 of SMM7 (see clause 2.14.1 and 2.14.3).

By far the largest number of variations arises under (a). In cases (a), (c) and (d), the architect is required to issue instructions but that is not necessary in cases (b), (e) and (f). All such instructions must be in writing. Consequently, although the architect may order any variation that he is empowered to issue under the contract conditions, the contractor is bound to carry out such an order (subject to certain exceptions – see later section) only if it is in the form of a written instruction issued by the architect. Therefore, if the contractor carries out any variation involving him in additional expense, and the variation has not been the subject of a written Architect's instructions (AIs), he runs the risk of being unable to recover the extra costs.

The contractor is safeguarded, however, in the event of instructions being given orally. Under clause 3.12.2 of the JCT Form, an oral instruction 'shall be confirmed in writing by the contractor to the architect within seven days'. The architect has the opportunity during seven days following receipt of that confirmation to dissent, otherwise the instruction becomes effective immediately. Failing such action, the architect may issue a confirmation at any later time but prior to the issue of the Final Certificate (clause 3.12.4). Where oral instructions have allegedly been given and complied with but not subsequently confirmed, some surveyors ask the architect to issue confirmation, if he is willing to do so.

The architect may thus issue a written instruction which sanctions a variation already carried out in compliance with oral instructions. Clause 3.14.4 of the JCT Form goes further, enabling the architect to sanction subsequently any variation which a contractor has already carried out not in pursuance of an AI, oral or written.

On contracts where there is a clerk of works, any directions which he may properly give to the contractor (whether orally or in writing) become effective only if confirmed in writing by the architect within two working days (clause 3.4).

It is necessary, therefore, for the surveyor to satisfy himself, before giving financial effect to any variation, that written instructions have been given by the architect, ordering or sanctioning the variation, or if the contractor claims to have confirmed oral instructions, that the architect does not dissent from such confirmation. The surveyor has no authority to adjust the Contract Sum in respect of a variation which has not been put into the form of a written AI or been otherwise confirmed in writing (see comment on p. 5).

Contractor's compliance

The contractor should be aware of the procedure that allows him to question the validity of an AI and the situations where his general obligation to comply with an AI may be modified. In the first instance an architect may only issue an instruction where the contract conditions expressly allow him to do so. If the contractor is of the opinion that an architect has issued an 'invalid' AI he has the right to request the architect to identify, in writing, the contract condition that empowers the architect to issue the AI in question (see clause 3.13). If the contractor is not satisfied with the architect's response he may refer the issue to adjudication, otherwise the instruction will be deemed to be valid. Where an architect issues a clause 5.1.2 instruction, e.g., an instruction to vary working hours or an alteration to the site access then the contractor need not comply with this instruction as long as he provides the architect with a reasonable objection in writing. The reason for this procedure is that a clause 5.1.2 instruction could have such a significant impact on site operations that it was thought necessary to build in a procedure to allow the contractor to protect his position. Similarly, if the architect issues an instruction which the contractor considers may adversely affect the design of the Contractor Designed Portion, the contractor need not comply with the instruction but must within 7 days of its receipt write and inform the architect of the perceived problems. However, if the architect subsequently confirms the instruction then the contractor is obliged to comply. Finally if the architect issues a clause 5.3.1 instruction, i.e., asking the contractor to provide a Schedule 2 Quotation, the contractor is not to comply with the variation work until the architect issues him with a confirmed acceptance of the quotation or issues an instruction under clause 5.3.2 for the work to proceed as a normal variation instruction. If the contractor is not willing to provide a Schedule 2 Quotation he must notify the architect within 7 days of the instruction being issued.

The contract conditions do not specify a set number of days by which a contractor is to comply with an instruction but merely requires the contractor to comply forthwith (see clause 3.10) which may be interpreted to mean that the contractor is expected to respond promptly to any instructions issued by the architect. If a contractor appears to be ignoring an instruction or refuses to comply with an instruction the architect may resolve the problem through the procedures in clause 3.11, i.e., the architect may send the contractor a compliance notice asking him to comply with a previously issued instruction. If after 7 days from receiving the compliance notice the contractor still fails to comply with the instruction the employer may bring in outside contractors to carry out work in connection with the instruction. The contractor will be liable for the additional costs incurred by the employer (refer to *Bath and North East Somerset District Council v Mowlem, 2004* for an example of this procedure being used).

It should be appreciated that not all AIs necessarily constitute or contain variations to the contract. Some may be explanatory, others may direct how provisional sums are to be expended, yet others may instruct the contractor to remove from the site materials not in accordance with the contract. None of these AIs may constitute a variation to the Works.

In some instances a variation may come into existence without the issue of an AI, e.g., a variation arising under clause 2.14.1 of the JCT Form, as listed in (e) above. This is where an error or omission has occurred in the compilation by the surveyor of the bills of quantities upon which a contract sum has been based. In such circumstances the error or omission must be corrected and 'shall be treated as a variation'. It should be noted that this does *not* relate to errors made by the contractor when pricing bills of quantities (see p. 91).

Measuring variations

Few, if any, variations have no effect on the Contract Sum. In order, therefore, to adjust the Contract Sum adequately and satisfactorily within the Conditions of the contract, the quantity surveyor (and the contractor's surveyor, too, if he wishes) must ascertain the net extra cost or net saving involved. To do that, the affected work will have to be measured, except in those instances where the method of valuation makes it unnecessary, as discussed below. This means that the work originally designed, which will not be required as a consequence of the variation, must be identified, isolated and valued, and that which will be required to replace the omitted work must be measured and valued. Such measurement may be done from drawings or, if drawings are not available, by physically measuring the substituted work on the site after it has been carried out.

This same procedure will be necessary whether the contract is based upon bills of firm quantities or on specification and drawings only. In the case of contracts based upon bills of approximate quantities and those based on a schedule of rates, the whole of the Works will have to be measured as described above. In all cases the valuing will be done in the manner discussed in the section below headed 'Valuing variations'.

It should be appreciated that in the case of 'without quantities' contracts, the work omitted will have to be measured in detail, as well as the substituted work, there being no simple procedure of identifying bill items affected, as in the case of 'with quantities' contracts. Where there are bills, it will often be necessary to refer back to the original dimension sheets in order to identify exactly what was included in the bills for the work which has been varied. If the items affected constitute complete bill items, they can be identified by their item references either as a single item or a group of items, e.g., 'item 132B' or 'items 132B–G'. If a variety of items is involved, many of which are not complete bill items, the omissions can be identified by reference to the dimension column numbers, e.g., 'original dimensions, X–X cols. 95–103', the letter 'X' being written on the dimension sheets to indicate the commencement and end of the omitted work. The measurement of AIs Nos. 8 and 48 in the examples at the end of the chapter illustrates the alternative procedures.

The quantity surveyor has the duty laid upon him by clause 5.4 of the JCT Form, to provide the opportunity for the contractor to be present when measurements of variations are taken. Although the clauses do not specifically refer solely to measurements on site, in practice the provision is not normally applied to measurement from drawings done in the office. It does mean, however, that the surveyor must give the contractor's surveyor sufficient notice for him to be able to make arrangements to attend on site when measuring is to be done there.

When variations are being measured this must be carried out using the same rules as used for the preparation of the contract bills (see clause 5.6.3.1), which in most cases will be the current Standard Method of Measurement of Building Works produced by the Royal Institution of Chartered Surveyors and the Construction Confederation. When measuring on site, it should be remembered that it is intended by the Committee that prepared the Standard Method of Measurement that the rules of measurement in that document should apply equally to measurement on site and to measurement from drawings.[1] This is important, for example, when measuring excavation involving working space and earthwork support. The SMM requires these items to be measured, even though the surveyor doing site measuring may know that they have not actually been done. The dimensions of the excavation should be those which are the minimum for the required construction, measured in accordance with the rules of the SMM, notwithstanding that the contractor may have excavated wider to avoid the use of earthwork supports or because the excavator bucket available was wider than needed. So, too, the lengths of timbers whose ends are not cut square (e.g., rafters) should be measured by their extreme lengths and inflexible floor tiles (e.g., quarry tiles) should be measured between skirtings, particularly where the latter are coved, and not *between* wall faces.

Measurements taken on site are normally entered in bound dimension books, which consist of pages with dimension column rulings similar to 'cut and shuffle' slips, or to traditional dimension paper but with only one set of columns instead of two. Alternatively, one may use normal dimension paper on a clipboard which, although not so convenient to handle as a dimension book, does give the advantage of having all the measurements, whether taken on site or in the office, on the same size sheets.

Another way of dealing with both the measurements and the working-up involved, is to use A4 size paper ruled with dimension columns on the left and quantity and money columns on the right – i.e., similar to estimating paper. Thus, all the information – measurements, billing and pricing – is presented on the same sheets (see AIs Nos. 15, 23, 28 and 41 in the examples on p. 145).

Variation accounts

When the measurement has been done, the dimensions have to be worked up into variation accounts, which, in compliance with clause 4.5.2.2 of the JCT Form, should be completed and priced and a copy supplied to the contractor not later than 9 months after Practical Completion. In practice, this may not always be achieved. Nevertheless, the surveyor should do his best to complete the variation accounts by the expiry of the period stated in the contract as the Employer is in breach of contract if he does not do so.

The format in which the variation accounts are drawn up varies between one firm of surveyors and another. Some firms bill omissions and additions under each AI separately. Others combine all the omissions together and all the additions together, each group being classified under appropriate 'Work Section' headings. The merit of the former method is that the cost of each variation is apparent, together with separate totals of omissions and additions. The main disadvantage is that the variation accounts are almost certain to form a thicker document due to the repetition of items in different variations and also because of the extra space taken up with separate totals. The second method results in less items and a more compact document which takes less time to prepare and price. If all that is required is the amount of the net extra or saving, then the format would seem to be adequate. If details of the costs of the individual variations are needed or desired, then the former method of presentation will probably be necessary. Apart from that, the format may be dictated by custom in individual surveying organizations.

Valuing variations

There are several ways of valuing variations, the choice in a particular case being that which is appropriate to the circumstances. They are:

(a) by the inclusion in the variation accounts of a lump sum in accordance with a quotation submitted by the contractor and accepted by the architect, e.g., a Schedule 2 Quotation;
(b) by pricing measured items in the variation accounts;
(c) by ascertaining the total prime cost of additional work and applying appropriate percentage additions, e.g., a dayworks sheet.

When a standard variation is issued the contract conditions require the employer and contractor to agree the value of the work executed (see clause 5.2.1). However, the contract conditions provide no advice on how this procedure is to operate, e.g., no one party is given responsibility for preparing a pricing document in the first instance, there are no rules or guidelines explaining how the prices should be determined, there is no timetable advising when meetings should be arranged, how long the parties have to consider price proposals before they accept or reject them or what happens if both parties fail to operate the procedure – how long would it be before this was deemed a failure to agree? All this procedure that was present in the 1998 edition of the contract has now been removed.

It is likely that few employers would have the in-house expertise to deal with the valuation of variations and in practice it is likely that the employer would delegate this task to his quantity surveyor. Therefore, at the early stage of the contract it would be advisable to inform the contractor of the party who is to represent the employer when agreeing the value of variations and reaching agreement on how the procedure should operate. Where a quantity surveyor acts for the employer under this procedure he is not bound to comply with the Valuation Rules in clauses 5.6 to 5.10, he is acting on behalf of the employer with the objective of trying to agree a valuation with the contractor – by whatever means. Although the employer and contractor are not bound by the valuation rules at this stage it is likely that they would be referred to in negotiations as a basis of determining what is a reasonable valuation.

If the employer and contractor are unable to agree the value of a variation then the quantity surveyor will normally take over the responsibility for valuing the work, unless the employer and contractor agree on an alternative approach, e.g., for a complex variation of a specialist nature the two parties may agree to refer the valuation to an independent expert.

Where the employer and contractor have been unable to reach an agreement on the valuation of a variation and the quantity surveyor has taken on this responsibility he must carry out the task in accordance with the Valuation Rules (clauses 5.6 to 5.10). In this situation the quantity surveyor has the right to decide which is the appropriate method and means of valuing a variation. Therefore, he is not bound to accept a statement on an AI to the effect that a variation is 'to be carried out on daywork' or 'to be paid as daywork'. In any case, such a statement may be contrary to the provisions of the JCT Form.

The alternative valuation methods listed above merit further discussion as follows.

Contractor's accepted quotation

Where the architect provides adequate information in an AI, clause 5.3.1 of the JCT Form makes provision for the contractor, to submit to the surveyor a quotation (known as a 'Schedule 2 Quotation') for carrying out the specified work. To ensure the contractor receives adequate information, the JCT advises that the AI should provide information in a similar format to the documentation provided at the tender stage, i.e., drawings and/or bill of quantities and/or a specification. The contractor is not obliged to provide a Schedule 2 Quotation as long as he gives the architect a written notice within 7 days of receiving the instruction (the 7-day period may be extended with the agreement of the employer and contractor).

The Schedule 2 Quotation must show separately and in sufficient detail:

1 the value of the adjustment to the Contract Sum, including the effect on any other work (e.g., a change in the conditions in which other work would have to be subsequently carried out, omitted work and adjustment of preliminary items) with references and supporting calculations based on rates and prices in the Contract Bills as appropriate;
2 any adjustment of time required for completion of the Works;
3 the amount to be paid in lieu of the value of direct loss and/or expense under clause 4.23;
4 a fair and reasonable sum for preparing the Schedule 2 Quotation;
5 where requested by the architect, the contractor is to provide information on the additional resources that may be required to carry out the work and a method statement explaining how the work is to be executed.

For each of the above sections, the contractor must provide supporting information which includes sufficient detail to allow the quotation to be properly evaluated by the Employer or his representative.

The Employer has seven days from receipt by the surveyor of the Schedule 2 Quotation in which to accept it in writing (this seven-day period may be extended with the agreement of the employer and contractor). If so accepted, the architect must immediately confirm the acceptance in writing to the contractor. In the 'Confirmed Acceptance' the architect is to state:

1 that the contractor is to carry out the valuation;
2 the consequential adjustment to the Contract Sum which is to include the amounts allowed for by the contractor for any direct loss and/or expense and the cost of preparing the quotation;
3 a revised Completion Date, where the contractor has stated that an adjustment of time is required.

This is an attractive way of valuing a variation because it eliminates the need for time-consuming measurement and pricing. However, it contains the conditions for inflated prices: no competition, and the liability that the quotation will be insufficiently scrutinized due to pressure of time and also the architect's understandable anxiety to avoid delay. But provided the surveyor makes an approximate estimate of *net* extra cost (i.e., taking account of omissions as well as additions), and the contractor's price bears a reasonable relationship to the estimate, then the danger will be minimized.

If the contractor does not agree to provide a Schedule 2 Quotation when so requested in an AI, the variation work is not to be carried out; unless the architect issues a further instruction for the variation to be executed and valued by a Valuation. This subsequent AI indicates that the work will now be valued by the quantity surveyor in accordance with the valuation rules clauses 5.6 to 5.10, and the employer and contractor will not be required to attempt to reach a prior agreement on value as clause 5.2.

If the Employer has not accepted a Schedule 2 Quotation by the end of the seven days allowed for acceptance, the architect must:

1 instruct the contractor that the variation is to be carried out and valued under the valuation rules in clauses 5.6 to 5.10; or
2 instruct that the variation is not to be carried out.

In either case, a 'fair and reasonable amount' is to be paid to the contractor for preparing the Schedule 2 Quotation as long as it is evident that the quotation presented by the contractor had itself been prepared on a fair and reasonable basis. This amount should have been separately identified in the contractor's quotation. However, if it is evident that the contractor had grossly exaggerated his quotation the employer would be within his rights not to make a payment for the preparation of the quotation.

It should be noted that clause 5.3 does not place any limit on the scope or the value of the work included in the variation which is the subject of the Schedule 2 Quotation. It follows that the architect can, within the Conditions of Contract, vary the works to include major changes in the scope and value of the Contract.

Pricing measured items

As previously stated where the contract conditions require the employer and contractor to try and agree the value of a variation (see clause 5.2) they are not bound by any specific pricing rules, they are free to use whatever means they like to try and reach an agreement. However, where an agreement cannot be reached and the task is passed to the quantity surveyor the variation must now be valued in accordance with the contract conditions. The JCT Form in clauses 5.6 to 5.10 is explicit about the way measured items (omissions as well as additions) should be priced. The prices (in particular, unit rates) in the bills or schedule of rates, as the case may be, are to be used. This is subject to the proviso that the character of the work and the conditions of its execution are similar to that in the project as originally envisaged. This covers the pricing of all (or most) omissions and much of the additions (although some omission items in 'without quantities' contracts may not coincide with items in the submitted schedule of rates).

Other items will bear no relationship or comparison at all with items in the contract bills, or may not be carried out under similar conditions to apparently similar items in the bills. In these a variation should be valued by using 'fair rates and prices'. This means that a unit rate may have to be built up from the prime cost of the necessary materials and using labour constants valued at 'all-in' labour rates and with any appropriate allowances for plant and with additions for overheads and profit. If an item is a common one or of small total value, the surveyor's knowledge of prices will usually enable him to put a fair price to it without the necessity for a detailed synthesis on the lines just described.

In between the two foregoing groups of items, there is a third group in which the items are not exactly the same as but bear a fairly close resemblance to items in the bills or schedule of rates. These should be priced *on the basis* of the prices of the comparable items. That means in effect, that the pricing of such items is to be done, as far as possible, at the same general level of prices as is contained in the contract documents. Again, where an item of work is common or relatively inexpensive, the surveyor will usually be able to 'assess' a comparable rate to the similar bill or schedule item. Where the item is an expensive one in total cost terms, or where there is any dispute, it may be necessary to analyse the contract rate and then to synthesize the new rate on the basis of the analysis. In any case, the student should be able to do price analysis and synthesis and the examples at the end of the chapter will illustrate the process.

Perhaps it should be emphasized that, in practice, it is only necessary for relatively few items in variation accounts to be priced by a process of analysis and synthesis. It should be realized, however, that, in theory, the same process is being carried out when a comparable rate is 'assessed'. As in many fields, experience enables one to short-circuit longer procedures, but this can only be done safely once the underlying principles and procedures have been fully understood and appreciated. Then (to give a simple illustration), the student will know that a 100-mm-thick concrete slab should not be priced at half the price of a 200 mm one to the same specification, even without the experience to be able immediately to suggest a reasonable price for the thinner slab.

Prime cost plus percentage additions

This method of valuing variations is commonly known as *daywork* and bears a close resemblance to the valuation of work under the prime cost or cost reimbursement type of contract (see p. 23). Daywork is a method of payment by the reimbursement to the contractor of the prime cost of all materials, labour and plant used in the carrying out of the work, with a percentage addition to the total cost of each of those groups for overheads and profit.

Daywork is subject to the same objection as that levied against prime cost contracts, namely, that its use generally results in higher costs to the Employer than when a 'measure and value' basis is used. Not surprisingly, therefore, daywork is favoured by contractors and deprecated by the Employer's professional advisers.

This generally acknowledged characteristic of daywork is the reason for the restriction of its use in the Standard Forms to situations where 'work cannot properly be valued by measurement'. This provision should be strictly observed in the Employer's interest. If work is capable of adequate measurement, it can usually also be priced using unit rates determined as described under *Pricing measured items* above.

Such measurement must be capable of being 'properly' done, which would exclude, for example, measuring some alteration or adaptation works by a single 'number item' with a long and complicated description, merely to avoid paying the contractor daywork. In any case, to price such an item would probably involve the surveyor in asking the contractor how long had been spent on the work, what materials and how much of them had been used, and what plant – which comes down to something very much like daywork in the end!

The surveyor should recognize and accept that the contractor has the right to be paid a daywork basis where appropriate and that the Employer has the right to pay on a measured basis where that can reasonably be done.

As with prime cost contracts, it is necessary to have a precise definition of what is included in the prime cost in each of its three divisions – materials, labour and plant. The JCT Form refers to the document *Definition of Prime Cost of Daywork carried out under a Building Contract*[2] as the basis to be used for the calculation of the prime cost, or, in the case of specialist work, a comparable document appropriate to that work, if one exists. For example, separate daywork definitions exist for electrical work as well as heating and ventilation work.

The respective percentages to be added to the totals of each section of the prime cost will be those included in the contract document, i.e., bills of quantities or schedule of rates. If by some oversight, no percentages were included, then they would have to be agreed between the surveyor and the contractor's surveyor.

The contractor is required to submit 'vouchers' (commonly called 'daywork sheets') not later than the end of the week following that in which the daywork was completed, giving full details of labour, materials and plant used in carrying it out. These vouchers should be verified by the architect or his authorized representative. If there is a clerk of works, he can usually be relied upon to do the verifying. The surveyor may be prepared to accept unverified daywork sheets where he is able to satisfy himself that the claim is reasonable, but to do so is not strictly within the terms of the Standard Forms. If verified sheets are accepted and the hours and quantities shown on them are considered to be excessive, some surveyors may consider reducing them to reasonable amounts. Whether a surveyor is entitled to act in such a manner is open to question.[3] If the authenticity of a daywork is to be challenged this should be done by the architect at the time of verification (*JDM Accord v DEFRA, 2004, 93 ConLR 133*).

It should be noted that a signature denoting verification by the architect or his representative relates only to the materials used and the time spent[4] and not to the correctness or otherwise of any prices included. The proper valuation is for the surveyor to determine.

There are difficulties associated with verifying daywork vouchers but as they are problems for the architect rather than the quantity surveyor, it is not considered necessary to discuss them here; the reader is referred to a relevant article in *Building*.[5]

The rates and prices used for valuing daywork should be those current at the time the work was carried out, not those current at the date of tender. This is so that the application of the respective percentage additions for overheads and profit will be to the total costs. If the rates and prices current at the date of tender were used, the total payment for the daywork would be less (assuming some increases in costs since the tender date). The difference would not be made up by the reimbursement of such increases under the fluctuations provisions of the contract, even if such provisions were included. It is also because rates and prices current at the time the daywork was carried out are used for valuing it, that the value of such work is excluded from the provisions relating to fluctuations in clause 4.21 of the JCT Form.[6]

The 'Definition of Prime Cost of Daywork' provides the rules that determine what may or may not be claimed within a daywork. A summary of the key rules is provided below.

Labour

The time that may be allowed for labour on a daywork sheet is 'the time spent by operatives directly engaged on daywork, including those operating mechanical plant and transport and erecting and dismantling other plant'. If any overtime is worked, only the actual hours worked may be claimed and not the enhanced rate (e.g., work paid for at time-and-a-half: 4 hours actually worked although operatives paid 6 hours, only 4 hours may be claimed) unless the overtime working is specifically ordered by the architect and the contractor and employer had previously entered into a written agreement that such additional overtime costs may be recovered. Supervisory staff such as foremen, gangers, etc., may only be included on a daywork sheet when manually employed on the variation work, when they are to be priced at rates appropriate for their trade.

The rates of labour used in valuing daywork should be the standard hourly rates based on the industry agreed working rule agreement, irrespective of the actual rates paid. Where rates above the basic rates are paid, the excess comes out of the percentage addition the contractor is entitled to include on the daywork sheet. The method of computing the standard hourly base rates for use in pricing dayworks is illustrated on p. 8 of the *Definition of Prime Cost of Daywork*, and a worked example is provided in Figure 7.1. It can be a complex task tracking down all the up to date information necessary for the calculation of the standard hourly base rate. To make this task easier, surveyors can usually obtain the current standard hourly base rates from the Building Cost Information Service (BCIS) or relevant trade organizations.

Materials and goods

The cost of materials on a daywork is to be the invoiced cost after trade discounts have been deducted. If there is a cash discount shown on an invoice (i.e., an allowance for prompt payment) this is not deducted unless it is in excess of 5%, when it is to be adjusted to allow the contractor 5% only. The cost of materials obtained from suppliers, etc., is to include the cost of delivery to site where this is applicable. If a contractor uses old materials from his yard for a daywork they are not priced at the original invoice rates but are to be charged at current market prices plus any appropriate handling charges in getting the materials to site. Any VAT that is shown on a supplier's invoice is not to be included in a daywork.

Plant

Plant is to be charged at the rates provided for in the contract documents. Most tender documents make allowance for the use of the *Schedule of Basic Plant Charges* prepared by the BCIS.[7] However, its use is not mandatory and any schedule or system may be adopted for use in the tender documents. The *Schedule of Basic Plant Charges* is published on an occasional basis, which means that the level of prices in the current edition is often out of date. If that is the case, then the rates quoted will require updating. Allowing the contractor to quote a percentage addition on the schedule of rates would achieve this. Alternatively the same result may be obtained by including an appropriate allowance within the percentage addition allowed on the plant costs on the daywork sheet (see Section 6.1 (q) from the Daywork Definition). Provision for both the use of the *Schedule* and for updating the rates ought to have been made in the contract bills or schedule of rates, as the case may be.

An example of the above calculation is provided below, and is based upon the rates relevant to a craft operative and a general operative under the Construction Industry Joint Council Working Rule Agreement. Rates and levies are applicable as at 28 June 2004.

The calculation for the number of 'standard working hours' per year is based on 52 weeks at 39 hours (the normal working week), less annual holidays (21 days) and public holidays (8 days) which equates to 29 days or 226 hours (23 days @ 8 hrs and 6 days @ 7 hrs).

Agreed rates from 28 June 2004		Craft operative	General operative
Weekly rate		320.58	241.02
Hourly rate		8.22	6.18
Supplemental payment – where applicable			
Guaranteed annual earnings: standard hours 1802 × hourly rate		14,812.44	11,136.36
Employer's national insurance @ 12.8% on earnings above £91 per week* (based on 46.2 weeks)		1,357.64	887.16
Holiday payments 226 hrs × hourly rate		1,857.72	1,396.68
Benefit scheme† 52 weeks @ £10		520.00	520.00
CITB levy 0.5% of	16,670.16	83.35	
	12,533.04		62.67
Annual prime cost of labour		18,631.15	14,002.87
Hourly rate (divide by 1802)		10.34	7.77

*The earnings threshold before the employer becomes liable for National Insurance is £91.

† Under the working rule agreement the amount the employer has to contribute to the pension scheme ranges from £2.51 to £10 per week depending upon an employee's contribution.

Having calculated the nominal annual cost of employing a specific grade of operative, this figure is then divided by the number of hours an operative would be on site for a year, using the standard working hours agreed between the unions and employers. The result will be an hourly base rate which may be used in the daywork to value the cost of labour.

Figure 7.1 Example of daywork labour calculation

Where the Schedule of Basic Plant Charges is being used it should be noted that in 'Explanatory Note 3', it states that the rates apply to plant, whether hired or owned, that is already on site. Also Note 5 states that the time allowed should be for when the plant is actually engaged in daywork. The situation is a little more complicated when plant has to be specifically hired for a daywork. The advice provided by Note 7 states '... daywork shall be settled at prices which are reasonably related to the rates in the Schedule having regard to any overall adjustment quoted by the Contractor ...'. It is not always clear how this should be interpreted. Although the schedule of rates applies to hired plant already on site it is necessary to understand why the same rates should not be equally applied to plant specifically hired in. Because the plant is specifically hired in for the daywork the cost of transport to and from site is recoverable by the contractor, in accordance with note 5.2.a) from the 'Definition of a Prime Cost of Daywork'. Also the time allowed for such plant is the time it is employed on the daywork, which should include erection and dismantling as well as periods of time when it may be standing idle during the progress of the daywork.

To take account of the above situation it may be advisable to have two separate plant items included in the daywork section of the bill of quantities (see example in Figure 7.2). This allows the contractor to quote a percentage addition for plant already on site, and in this he would obviously include a percentage to lift the prices to present-day levels plus incidental costs, overheads and profit. In the second instance, where plant is specifically hired for the daywork, the cost would be determined and in the percentage addition the contractor would be seeking to recover mainly profit, overheads and other incidental costs.

Unless stated otherwise the rates in the plant schedule normally include all allowances for fuel, maintenance, licences and insurances. In many instances the rates in the plant schedule make no allowance for the cost of driver or attendants although there are exceptions, e.g., mobile and tower cranes, lorries and dumper trucks. Within the plant section the cost of non-mechanical hand tools and scaffolding, staging and trestles, etc., that are already erected in position, is not recoverable. Scaffolding and such like may only be included when it has been specifically erected for the daywork.

Figure 7.8 (p. 145) gives an example of a daywork sheet and Figure 7.7 (p. 143) shows the same information converted into a daywork account for incorporation into a variation account.

	Dayworks		
	For the valuation of variations relating to the execution of additional or substituted work which cannot properly be valued by measurement, the contractor will be allowed to charge daywork in accordance with the provisions of clause 5.7 of the Conditions of Contract.		
	The basis of charging will therefore be the Prime Cost of such calculated in accordance with the 'Definition of Prime Cost of Daywork carried out under a Building Contract' which was current at Base Date, together with percentage additions to each section of the prime cost at the rates set out below by the contractor.		
	Provide the following Provisional Sums and add thereto the percentage addition required for the incidental costs, overheads and profit.		
	For the Prime Cost of:		
A	Labour		1,500.00
B	**Add** Percentage addition required		
C	Materials and Goods		1,000.00
D	**Add** percentage addition required		
E	Plant (where charged at the rates referred to in paragraph [a] of the note below)		750.00
F	**Add** percentage addition required		
G	Plant (where charged at the current rates as referred to in paragraph [b] of the note below)		200.00
H	**Add** percentage addition		
	Note: In pricing the percentage addition required in respect of Plant the contractor should note that the rates allowed for the individual items of plant will be either:		
	[a] the rates contained in the 'Schedule of Basic Plant Charges' for use in connection with Daywork under a Building Contract issued by the RICS and which was current at Base Date or,		
	[b] in those cases where, for any reason the latter Schedule is not applicable at rates current at the time the work is executed.		
	Carried to collection		

Figure 7.1 Example of a bill of quantities daywork section

Additional expense arising from variations

There is a possibility that the contractor may not always be adequately remunerated for what he is required to do in regard to variations under the terms of the Standard Forms. For example, a variation which increased the width of a number of windows might be ordered. The timing of the order might result in the contractor having bought lintels for the window openings which had been delivered before the variation was ordered, and which, in consequence became surplus to requirements. The normal measurement of the omissions and additions involved would result in the omission of the shorter lintels and the addition of longer ones and no account would be taken of the surplus lintels. The possibility of additional expense arising from variations is recognized by the JCT Form in clause 5.10.1. Where the quantity surveyor is valuing the variation it may be advisable for the contractor to provide notification of this expense. If the notification is deemed to be justified, the surveyor must ascertain the amount involved, which then must be added to the Contract Sum. Any such amount should also be included in the next interim certificate issued under clause 4.9.

Use of erroneous rates when pricing variations

Sometimes, a rate against a measured item in bills of quantities or in a schedule of rates is inadvertently underpriced or overpriced by the contractor and the error may have escaped the notice of the surveyor when examining the tender document before the contract was let, so that nothing was done about it (see p. 91). If, subsequently, the item is involved in a variation, then a difficulty may arise in the use of the erroneous rate, which must be resolved.

Suppose, for example, that an item of 'two-coat plaster on walls' had been priced at £0.69 per m^2 in mistake for £6.90 per m^2. No problem need arise over omissions, as the omitted quantity should, without doubt, be priced at the bill rate. But suppose the additions involve a much larger quantity than that omitted. At what rate should the addition quantity be priced? Is it inequitable to the contractor to price it at the underpriced rate? Is the Employer taking unfair advantage of a genuine error on the contractor's part?

The answers to these questions should be sought by asking 'what does the contract say?' The JCT Form in clause 5.6.1.1 clearly states that bill rates shall be used, provided the work is of similar character and executed under similar conditions. There is no authority therefore for the surveyor to use a different rate, so long as the work is 'similar' in those respects. As in other aspects of contractual agreements, what has been freely agreed by both parties is binding upon them, notwithstanding that a term of a contract may subsequently prove to be disadvantageous to one of the parties. It seems clear, therefore, that the contractor must stand the loss (or be allowed to gain the benefit, as the case may be) of an erroneously priced item that is involved in variations.

A compromise that has been suggested from time to time is that the erroneous rate should be applied only up to the limit of the quantity of the item in the contract bills to which it applies and that beyond that limit a fair rate should be substituted. This will appear to many people as an equitable solution although it must be said again that there is no direct authority in the JCT Form for so doing. The judgment in a case heard in the High Court in 1957, *Dudley Corporation v Parsons & Morris Ltd*, held that the total quantity of an item in a re-measurement should be priced at the erroneous rate in question, a decision which was upheld in the Court of Appeal in 1959. The more recent case of *Aldi Stores Ltd v Galliford (2000)* has confirmed and reinforced the above principles. This appears to settle the matter.

The question has been raised whether, in these circumstances, the contractor can justifiably claim reimbursement of the loss resulting from applying the erroneous rate to the additions quantity, under the provision of clause 5.10.1 of the JCT Form. Legal opinion that such a loss could not be recovered under clause 11(6) of the 1963 Edition of the JCT Form[8] would seem to apply also under clause 5.10.1 of the 2005 Edition.

Adjustment of Preliminaries on account of variations

This matter has at times been a cause of contention and disagreement. Some contractor's surveyors have argued that if the value of the job is increased by variations, then the Preliminaries should also be increased in proportion, on the grounds that, by and large, they are directly affected by the size and value of the contract.

On the other hand, some surveyors have contended that the Preliminaries should only be adjusted in very exceptional circumstances, i.e., when the size and scope of the job is significantly affected, as, for example, by the addition of another storey or another wing or block. Others have taken the intermediate position that if there has been an extension of time granted in respect of variations, then there may be grounds for adjusting part of the value of Preliminaries.

Editions of the JCT Form published before 1980 were of little or no help in providing a solution, as they made no specific reference to Preliminaries. However, the 1980 Edition and subsequent editions of the JCT Form do provide specific advice on the matter (see clause 5.6.3.3). This requires that, when valuing additional, omitted or substituted work, allowance is to be made for any addition to or reduction of preliminary items. It should be noted, however, that the provision relates only to work 'which can properly be valued by measurement' and so excludes other work, such as daywork. Also excluded is work pursuant to an AI for expenditure of a provisional sum for 'defined work', as long as the description for the defined provisional sum was accurate (see clause 2.14.1).

If the surveyor has to ascertain the value of any addition to or reduction of Preliminaries, how may he do so? He will first have to satisfy himself as to the proper total value of the Preliminaries. If there are bills of quantities and the Preliminaries are fully priced, the total value is apparent. If, however, they are not fully priced, an assessment of their full value will need to be made, because the JCT Form requires allowance to be made for additions to or reduction of the items. In order to make such allowance, the value of the items as originally intended needs to be known.

In the case of contracts on a 'without quantities' basis, the contract sum will need to be apportioned to its principal constituents, of which the Preliminaries form one, unless a Contract Sum Analysis has already been provided. When the total value of the Preliminaries has been determined, the surveyor must then ascertain the value of the individual items which comprise that part of the Contract Sum. Then he will be in a position to proceed with his valuation of any addition or reduction, as the case may be.

The pricing of Preliminary items by means of fixed charges and time-related charges was introduced into the *Standard Method of Measurement of Building Works* for the first time in the Seventh Edition (SMM7). It is intended that in the preparation of bills of quantities, etc., opportunity should be given for a fixed charge and a time-related charge to be included for all items of Employer's requirements (clauses A30–A37) and contractor's general cost items (clauses A40–A44). It is considered that this facilitates the adjustment of Preliminaries where this is required by clause 5.6.3.3 of the JCT Form when valuing variations. The fixed-charge element for each item can be analysed further for valuation purposes, into sums expended early and those expended late in the contract, as indicated in Table 12.2.

If an extension of time has been granted by the architect on account of variations, then those Preliminary items with a time-related content may be considered as ranking for a proportionate increase in that content, whilst the cost-related and fixed charge elements remain unchanged.

If no extension has been granted, there may still be justifiable grounds for adjusting the item prices. For example, a variation might result in plant being kept on site longer than otherwise envisaged or in additional supervision being required for the period during which the variation is being carried out. In such situations, the actual expense would need to be ascertained, which would require the assistance of the contractor in providing the necessary details.

It should be appreciated, however, that in no case should the total value of the Preliminaries be increased (or decreased) simply in proportion to the total value of variations. Any adjustment must be made with regard to the amount by which individual Preliminary items have been affected by variations. Although the SMM7 has made the task of assessing the value of Preliminaries slightly easier for the quantity surveyor by requiring the contractor to provide a breakdown of his time-related and fixed charges; it can still be a difficult task to realistically assess the value of additional preliminaries to be allocated to a variation. For example, a variation to the roof design may require the contractor to retain part of his access scaffold for a week longer than planned, and to erect additional scaffolding that had not been originally planned. The quantity surveyor should be aware that the contractor is entitled to these additional scaffolding costs, but reference to the Preliminaries section of the bill of quantities will be of limited value. The scaffolding in the Preliminaries bill will have been priced as two lump sums, i.e., the time-related charge and fixed charge. The quantity surveyor will not have access to the contractor's method statement and costings and will have little information as to how these lump sums were calculated, there will be no breakdown such as cost per square metre or cost per day. In these instances, the surveyor is going to have to fall back on experience and negotiate with the contractor.

Variations on contractor's designed portion work

Where a contractor has the responsibility for designing a portion of the works (i.e., CDP) it is still possible for the architect to issue an instruction resulting in a variation to the contractor's designed work (see clause 3.14.3). The value of the variation is to be agreed between the Employer and contractor or failing that, by the quantity surveyor.

A problem with attempting to agree the value of a variation on contractor's design work is the lack of detailed cost information available in the CDP Analysis. This analysis will only provide a simple cost breakdown of the designed work and it will not provide such detailed information as is readily available in a bill of quantities. If a quantity surveyor is required to value a CDP variation this must be carried out in accordance with clause 5.8. The procedures are very similar to the task of valuing a normal variation but because there are no bill rates to refer to the surveyor has to value the work on a basis that reflects the values set out in the CDP Analysis. This lack of detailed cost information makes it difficult for the Employer's surveyor to fully understand the cost implications of the work and may provide the contractor with the opportunity to try and maximize the value of the variation work. The surveyor will have to rely on his experience and knowledge of construction costs to try and ensure that any valuation is fair to both the Employer and the contractor. Finally, it is important to remember that the valuation must also take into account the cost of any associated design work of the contractor (see clause 5.8.1).

Examples of variations

The following table shows some of the variations contained in AIs issued for the contract of which particulars are given in Appendix A:

AI No.	Subject of variation
8	Willingdon Light Buff Facing Bricks in lieu of Hildon Tudor Brown Facing Bricks
15	The roof slabs to the flats increased from 125 mm thick to 175 mm and the main reinforcement increased from 12 mm to 16 mm diameter
23	Living room windows in flats to be 1800 × 1500 mm in lieu of 1800 × 1200 mm
28	Glazed screens to balconies to be glazed with toughened clear float glass with putty in lieu of clear float glass in beads
41	Window removed from existing store, opening in wall altered as necessary, provision of door and frame complete with deadlock, handles and closer
48	Omission of built-in bookcase units in living rooms

Figures 7.3–7.8 show the taking-off and billing for the variations. The work was billed direct from the taking-off in the case of AIs Nos. 8 and 48. The other AIs illustrate the use of 'estimating paper' ruling as an alternative format.

The following is a commentary on the examples.

AI No. 8 Facing bricks

This variation was necessitated by Hildon Brickworks Ltd ceasing the manufacture of Tudor Brown Facings. The *additions* items were all the same as the corresponding bill items apart from the change in bricks. As the latter were comparable in type and quality with those originally specified, the rates were adjusted merely to take account of the increased prime cost of the bricks, the value of the other materials and labour being unchanged. The price of the Hildon Tudor Brown Facings used by the contractor when pricing the bills was £270.50 per 1000 including delivery. The price of the Willingdon

Shops and flats, Thames St., Skinton

					£ p
	A.I. No 8				
	Willingdon Light Buff facings in lieu of Hildon Tudor Brown facings				
	Omissions				
A	Items 46A - 47A inclusive				113,739.41
B	Items 144F - 145C inclusive				24,967.19
	Total of omissions carried to summary				138,706.90
	Additions				
	Facework in Willingdon Light Buff facings and pointing in cement: lime: sand (1:1:6) mortar as the work p[roceeds				
C	Walls, 102 mm thick, facework one side p.r. item 46A	510	m^2	47.66	15,662.10
D	Ditto, 215 mm thick, ditto, p.r. item 46B	1272	m^2	78.91	15,662.10
E	Ditto, 215 mm thick, facework both sides, p.r. item 46C	12	m^2	89.53	15,662.10
F	Extra over, facework to reveals, 102 mm wide, p.r. item 46D	322	m	0.96	15,662.10
G	Ditto, Ditto, 215 mm wide, p.r. item 46E	16	m	0.96	15,662.10
H	Closing cavities, 50 mm wide, p.r. item 46F	42	m	4.86	15,662.10
J	Extra over, flush horizontal brick-on-end band, 215 mm wide, p.r. item 47A	218	m	15.14	15,662.10
K	Walls, 102 mm thick, facework both sides p.r. item 1446F	143	m^2	51.58	15,662.10
		Carried forward			136,959.54

Figure 7.3 AI No. 8

Shops and flats, Thames St., Skinton

	Description	Qty	Unit	Rate	£ p
	A.I. No. 8 (cont)				
	brought forward				136,959.54
	Facework in Willingdon Light Buff facings (cont)				
A	Walls, 215 mm thick, facework one side; p.r. item 144G	6	m²	78.91	473.46
B	Ditto, 215 mm thick, facework both side, p.r. item 144H	46	m²	89.53	4,118.38
C	Extra over, facework to reveals, 102 mm wide, p.r. item 145A	28	m	0.96	26.88
D	Ditto, ditto to 215 mm mide, p.r. item, 145D	11	m	0.96	10.56
E	Coping, brick-on-end, 215 mm wide horizontal, p.r. item 145C	182	m	15.50	2,821.00
	Total of additions carried to summary				144,409.82

Figure 7.3 continued

Shops and flats, Thames St., Skinton

				Description	Qty	Rate	£	p
		A. I. No. 15		Roof slab increased in thickness from 125 mm to 175 mm			£	p
				Omissions				
A		53 m3		Concrete slabs, thickness not exceeding 150 mm, reinforced, as item 36C	53 m3	109.85	5,822	05
B	293	12.00	3516.00	Bar reinforcement, 12 mm, straight, horizontal, length not exceeding 12 m, as item 38F	3.09 t	864.75	2,672	08
				Total of omissions carried to summary			8,494	13
				Additions				
C		74 m3		Concrete slabs, thickness 150 - 450 mm, reinforced, p. r. item 36C	74 m3	99.71	7,378	54
D	293	12.00	3516.00	Bar reinforcement, 16 mm, straight, horizontal, length not exceeding 12 m, p. r. item 38F	5.56 t	840.65	4,674	01
				Total of additions carried to summary			12,052	55

Figure 7.4 AI No. 15

Shops and flats, Thames St., Skinton

	Timesing	Dimension	Squaring	Description	Quantity	Rate	£	p
		A.I. No. 23		Larger windows in living rooms			£	p
				Omissions				
A	10/	1	10	Softwood window, 1800 x 1200, as D.J. Joinery Ltd's catalogue, ref. WM 1812N, as item 93C	10 No.	250.00	2,500	00
B	10/2/3/	0.50 0.35	10.50	Standard plain glass, panes area 0.15 - 4.00 m², as item 103B				
	10/3/	0.50 0.37	6.11					
			16.61		17 m²	25.98	441	66
C	10/	1.80 1.20	21.60	Two undercoats, one gloss finishing coat on ready-primed wood, glazed windows as item 108A	22 m²	6.37	140	14
				&				
D				Ditto external as item 109C	22 m²	6.40	140	80
E	10/	1.80 0.30	5.40	Walls, 102 mm thick, facework one side as item 1C, A.I. No. 8	5 m²	47.66	238	30
				&				
F				Ditto, 100 mm thick, lightweight concrete blocks, as item 42G	5 m²	24.21	121	05
				&				
G				Forming cavities in hollow walls as item 43B	5 m²	1.63	8	15
				carried forward			3,590	10

Figure 7.5 AI No. 23

Shops and flats, Thames St., Skinton

				Description	Qty	Rate	£	p
	A.I. No. 23 (cont)						£	p
				Omissions brought forward			3,590	10
A	10/	1.80 0.30	5.40	Two-coats plaster, blockwork walls as item 107D	5 m²	13.14	65	70
				&				
B				Emulsion paint, general surfaces of plaster, as item 107A	5 m²	3.54	17	70
				Total of omissions carried to summary			3,673	50
				Additions				
C	10/	1	10	Softwood window, 1800 x 1500 mm, as D. J. Joinery Ltd's catalogue, ref WM1815N, p.r. item 93C	10 No.	115.78	1,157	80
D	10/2/	0.30	6.00	Bedding frame and pointing one side	6 m	0.95	5	70
E	10/4/	0.55 0.37	8.14	Standard plain glass, panes area 0.15 - 4.00 m², as item 103B				
	10/2/4/	0.50 0.34	13.60					
			21.74		22 m²	25.98	571	56
F	10/	1.80 1.50	27.00	Two undercoats, one gloss finishing coat on ready-primed wood, glazed windows as item 108H	27 m²	6.37	171	99
				&				
G				Ditto, external, as item 109C	27 m²	6.40	172	80
				carried forward			2,079	85

Figure 7.5 continued

Shops and flats, Thames St., Skinton							£	p
	A.I. No. 23 (cont)							
			Additions	brought forward			2,079	85
A	10/2/	0.30	6.00	Closing cavities, 50 mm wide, as item 43J	6 m	4.86	29	16
				&				
B				Extra over, facework to reveals, 102 mm wide, as item 1F, A.I. No 8	6 m	0.96	5	76
C	10/2/	0.10 0.30	0.60	Damp-proof course, vertical 102 mm wide, as item 44C	1 m²	13.27	13	27
D	10/2/	0.16	3.20	Two-coats plaster, blockwork walls, not exceeding 300 mm wide, as item 107H	3 m	6.98	20	94
E	10/2/	0.16 0.30	0.96	Emulsion paint, general surfaces of plaster, as item 107A	1 m²	3.54	3	54
				Total of additions carried to summary			2,152	52

Figure 7.5 continued

Shops and flats, Thames St., Skinton

	Timesing	Dimensions	Squaring	Description	Quantity	Rate	£	p
		A.I. No. 28		Glazing to balcony screen			£	p
				Omissions				
A	10/2/	0.55 0.53	5.83	Standard plain glass, panes area 0.15 - 4.00 m², glazing to wood with beads and glazing strips as item 103A				
	10/	0.47 0.45	2.12					
	10/2/	0.55 1.17	12.87					
	10/	0.47 1.08	5.08					
			25.90		26 m²	27.85	724	10
B	10/2/	2.16	43.20	Glazing beads, 10 x 30 mm, mahogany, as item 115J	161 m	2.16	347	76
	10/	1.84	18.40					
	10/2/	3.44	68.80	&				
C	10/	3.10	31.00	Glazing strip as item 104B	161 m	1.08	173	88
			161.40					
				Total of omissions carried to summary			1,245	74
				Additions				
D			25.90	Special glass, toughened, panes area 0.15 - 4.00 m², glazing to wood with sprigs and glazing compound, p.r item 103B	26 m²	41.30	1,073	80
				Total of additions carried to summary			1,073	80

Figure 7.6 AI No. 28

Shops and flats, Thames St., Skinton

						£	p
	A.I. No. 41		New door to existing Store				
			Omissions				
A	0.68 0.90	0.57	One undercoat, one gloss finishing coat on existing wood glazed window as item 108E	1 m²	6.07	6	07
			&				
B			Ditto, external, as item 109F	1 m²	6.10	6	10
			Total omissions carried to summary			12	17
			Additions				
C	1		Door, 838 x 1981 x 44 mm, exterior quality ply-faced, p. r. item 96C	1 No	67.30	67	30
			&				
D			Door frame, 63 x 88 mm, rebated, for door size 838 x 1981 mm, p.r. item 95D	1 No	82.24	82	24
E	2/ 3	6	Frame tie as item 43L	6 No	0.77	4	62
			838 1981 2/ 2819 5638 2/2/2/50 400 6038				
F	6.04		Bedding frame and pointing one side as item 98D	6 m	1.61	9	66
G	1½		Pair butts, 100 mm, steel, as item 114D	1½ No	4.30	6	45
			Carried forward			170	27

Figure 7.7 AI No. 41

Shops and flats, Thames St., Skinton

						£	p
	A.I. No. 41 (cont)						
			Additions				
			brought forward			170	27
A	1		Mortice deadlock as item 114M	1 No	21.85	21	85
			&				
B			Pull handle, 75 x 200 mm, as item 114H	1 No	8.45	8	45
			&				
C			Push plate, 75 x 200 mm	1 No	6.41	6	41
			&				
D			Door closer as item 114Q	1 No	53.32	53	32
E	0.88 2.03	1.79	One coat primer, two undercoats, one gloss finishing coat, wood general surfaces as item 108D	2 m²	5.26	10	52
			&				
F			Ditto, external as item 108K	2 m²	5.40	10	80
G	4.90	4.90	Ditto, wood general surfaces. Not exceeding 300 mm girth, as item 108G	5 m	2.00	10	00
			&				
H			Ditto, external, as item 109B	5 m	2.05	10	25
J			Daywork account as daywork sheet no. 6 dated 19 June 2006			1,084	25
			Total of additions carried to summary			1,386	12

Figure 7.7 continued

Beecon Ltd River Road, Skinton

Contract: Shops and Flats, Thames Rd, Skinton   Ref. AI no.: 41

Daywork sheet no.: 6   Week commencing 19 June 2006

Signed (Architect or Representative) *S Draw*   Date signed: 30 June 2006   Total daywork (£) **1084.25**

Brief description of work: Take out window from existing store. Alter opening as necessary to form door opening, provide and build in Catnic lintel. Provide and fix 838 × 1981 door and frame, fit deadlock, handles and closer

Operatives	Trade	Hours M T W T F S S	Total		Rate	£	Plant as RICS schedule	M T W T F S S	Total	Rate	£
H Jones	Bklayer	8 8 4	20		11.34	226.80	Kango 2500	4 2	6	0.90	5.40
T Prince	Joiner	8	~~8~~								
G Howie	Labr	8 4	20		8.53	170.60					
B Scott	Painter	1 2 3	~~6~~								
				Sub-total		397.40			Sub-total		5.40
				Percentage adjustment	150%	596.10			Percentage adjustment	40%	2.16
				Total		993.50			Total		7.56

Materials	Qnty	Unit	Rate	£	Other plant	M T W T F S S	Total	Rate	£
Facing bricks	60	no.	300/m	18.00					
Themalite blocks 450 × 225 × 100	18	no.	1.10						
Mortar	0.05	m^3	67.50	3.38					
Catnic cavity lintel 1500	1	no.	43.60	43.60					
102 damp-proof course	5	m	0.87	4.35					
838 × 1981 Ext qual ply-faced flush dr	~~1~~	~~no.~~							
Door frame set	~~1~~	~~no.~~							
Undercoat paint	~~0.25~~	~~ltr.~~							
Gloss paint	~~0.25~~	~~ltr.~~							
		Sub-total		69.33		Sub-total			
		Percentage adjustment	20%	13.87		Percentage adjustment			
		Total		83.19		Total			

Figure 7.8 Example of daywork sheet

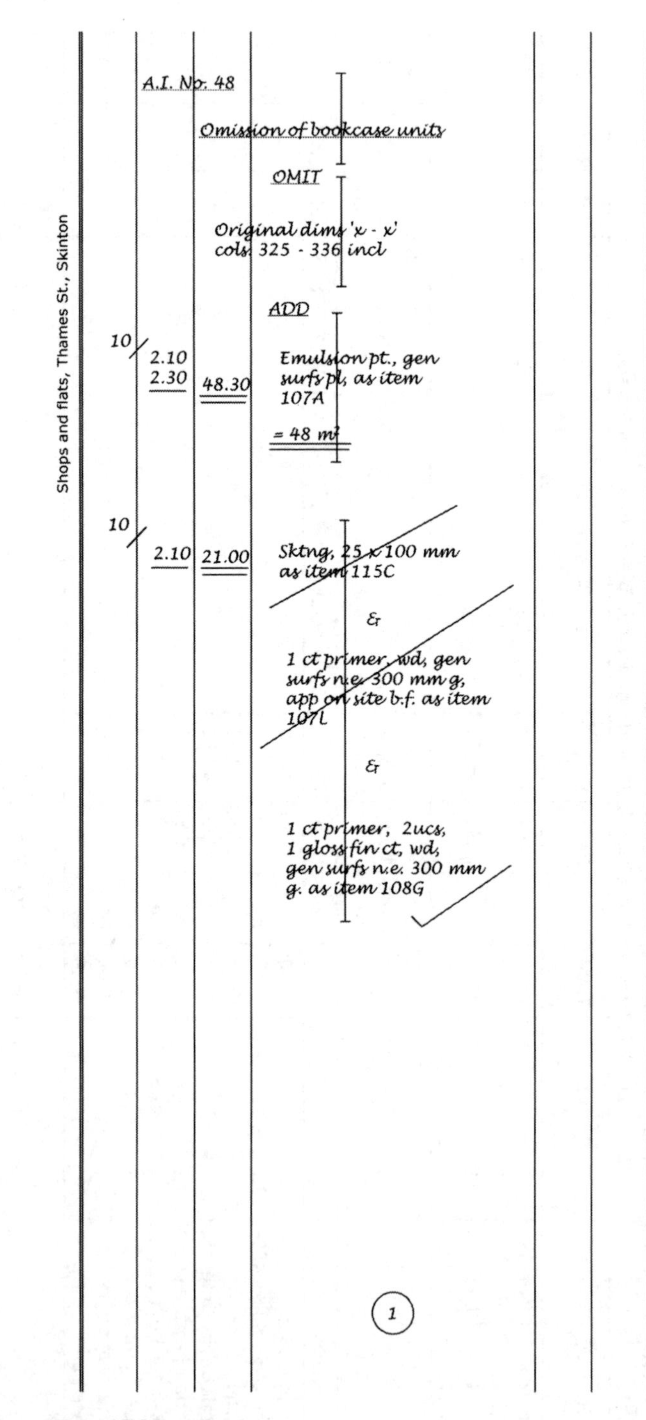

Figure 7.9 AI No. 48

Shops and flats, Thames St., Skinton

					£ p
	A.I. No 48				
	Omission of bookcase units				
	Omissions				
A	Items 110A - 110K inclusive				3,559.68
B	Item 112C				241.92
C	Item 112F				18.36
D	Two coats polyurethane clear finish wood, general surfaces, as item 107J	22	m^2	3.48	76.56
E	Ditto, not exceeding 300 mm girth, as item 107K	66	m	1.33	87.78
	Total omissions carried to summary				3,984.30
	Additions				
F	Skirtings, 25 x 100, softwood as item 115C	21	m	4.12	86.52
G	Emulsion paint, general surfaces of plaster as item 107A	48	m^2	3.54	169.92
H	One coat primer, wood general surfaces not exceeding 300 mm girth, applied on site prior to fixing, as item 107L	21	m	0.50	10.50
J	One coat primer, two undercoats, one gloss finishing coat wood general surfaces not exceeding 300 mm girth, as item 108G	21	m	2.00	42.00
	Total additions carried to summary				308.94

Figure 7.9 continued

Light Buff Facings, delivered, was £300 per 1000 and this price had remained unchanged since before tender date. The rates for the 102 and 215 mm walls items were calculated as shown below.

	£
Prime cost of Willingdon Light Buffs per 1000	300.00
Prime cost of Hildon Tudor Browns per 1000	270.50
Difference	29.50
102 mm walls	
63 bricks per m^2, Stretcher Bond,	
@ £29.50 per 1000	1.86
15% OHP*	0.28
	2.14
Bill rate, item 46A	45.52
Required rate	47.66
215 mm walls, facework one side	
85 bricks per m^2, Flemish Bond, @ £29.50 per 1000	2.51
15% OHP	0.38
	2.89
Bill rate, item 46B	76.02
Required rate	78.91

*The contractor stated, upon request, that the tender included 15% mark-up on net cost for overheads and profit.

AI No. 15 Roof slabs

This variation is shown using the format described on p. 117 in which the dimensions, billing and pricing are on the same sheet. The calculation of the *pro rata* rates for the additions was made as follows.

Concrete slab

The bill rate was first analysed to find the approximate rate used for the concrete at the mixer and this was then adjusted for the reduced labour input due to the increased thickness. It was considered that the increased size of reinforcing bars did not affect the placing of the concrete sufficiently to justify amendment of the labour content.

	£
Analysis of the bill rate	
Bill rate, item 36C	109.85
15% OHP (i.e., deduct 15/115)	14.33
	95.52
Labour placing, say 4.00 man/h @ £11.76*	47.04
Cost of concrete at mixer per m^3	48.48
Build-up of new rate	
Cost of concrete at mixer per m^3	48.48
Labour placing, say 3.25 man/h @ £11.76	38.22
	86.70
15% OHP	13.01
Required rate	99.71

*All-in labour rate calculated in accordance with the principles of the IOB *Code of Estimating Practice*.

Reinforcement

An increase in the diameter of reinforcing bars results in a lower cost per tonne of both the material and the labour content.

	£
Analysis of the bill rate	
Bill rate, item 30F, 12 mm bars, per tonne	864.75
15% OHP	112.79
	751.96
Cost of bars	480.59
Labour, bending and fixing	271.37
Build-up of new rate	
16 mm high tensile steel bars, per tonne	479.41
Labour, bending and fixing, say	251.59
	731.00
15% OHP	109.65
Required rate	840.65

*The balance of cost is reasonable for the required labour, thus indicating that the analysis of the bill rate is reasonable.

Formwork

No adjustment to the formwork rates is necessary. Although that to the edge of the roof is increased in height, it is still in the same height classification as that to the thinner slab, and the rate therefore remains the same (see SMM7, E20).

AI No. 23 Larger windows in living rooms in flats

Only the items affected by the variation were omitted from the original dimensions, i.e., the 'vertical' items. It should be noted that the deduction of walling for the larger windows can be taken-off either as 'deducts' items in the additions, or as omissions items. In the former case, as in the example, the results show negative quantities and so these items have been billed as omission items with positive quantities.

The windows item in the bills was priced at a rate of £250.00. This was analysed and a rate found for the larger windows as follows:

	£
Bill rate, item 64C	250.00
15% OHP	32.60
	217.40
P.C. of window, delivered	199.50
Labour, handling and fixing	17.90

The labour for the larger window was increased by 20% and added to the P.C. of the larger window, as follows:

	£
P.C. window, delivered	229.50
Labour, handling and fixing	21.48
	250.98
15% OHP	37.65
Required rate	288.63

The remaining rates required to price the additions items are either bill rates or those used in pricing items in AI No. 8. The latter are the brick facework items for which rates are now contained in the additions items of AI No. 8. The original bill rates for those items are, of course, no longer relevant.

AI No. 28 Glazing to balcony screens

In order to determine a rate for the toughened glazing, the bill rate was analysed and a new rate was then synthesized from the resulting data, thus:

	£
Bill rate, item 103A, per m^2	27.85
15% OHP	3.63
	24.22
4 mm float glass, including waste, say	21.38
Labour	2.84

The balance left for labour is reasonable, representing about 0.25 man/h (on the basis of an 'all-in' craftsman's rate of £11.76/h). So again, the data from the analysis may be used for building-up a new rate, as follows:

	£
4 mm toughened glass	32.10
Putty and glazing sprigs, say	0.40
Labour as before +20%	3.41
	35.91
15% OHP	5.39
Required rate	41.30

An alternative approach

It is not always possible for a surveyor to determine the prime cost of materials with a high degree of accuracy or to know with certainty the actual percentage mark up a contractor has used to recover his profit and overheads. In such instances the surveyor may adopt a different approach to determine a fair price for a variation or to check prices submitted by a contractor. Through experience or reference to pricing books a surveyor should be able to assess how the costs would normally be apportioned in a bill rate with regard to labour, materials and plant, and from this simple breakdown it should be possible to build a new rate suitable for the variation. The following example shows how this alternative approach may be used for AI No. 9.

The bill rate for item 103A, 4 mm float glass is £27.85. It takes a relatively short period of time to fix a pane of glass, which means the labour cost should be a minor part of the bill rate. It would not be unreasonable to assume a material to labour split of 85–15% (i.e., material £23.67 and labour £4.18). Toughened glass is approximately 40% more expensive than normal float glass, and glazing with sprigs and putty will take slightly longer than with glazing strip and beads, say an extra 20%. Therefore, a new rate would be:

4 mm toughened glass, say £23.67 plus 50%	35.51
Labour in glazing, say £4.18 plus 20%	5.02
Required rate	40.53

AI No. 41 New door to existing store

The omissions relate to the redecoration of the existing window, as provided for in the bills.

The contractor submitted Daywork Sheet No. 6, dated 30 June 2006, showing all the labour and materials expended. The surveyor decided, however, that the new joinery, ironmongery and painting were items that were properly measurable and priceable and should not therefore be valued as dayworks. The carpenter's and painter's time, as well as the relevant materials, have therefore been omitted from the daywork account.

The labour rates and materials prices used in valuing the daywork are those current at the dates when the work was done, i.e., 19–23 June 2006. The labour rates have been calculated in accordance with Section 3 of the *Definition of Prime Cost of Dayworks carried out under a Building Contract*. The respective percentage additions are those tendered by the contractor in the bills (see Appendix A, p. 385).

An addition item for which there was no directly comparable bill item was the item for the push plate. A rate was determined using fair rates and prices:

Steel push plate 75 × 200; Boltons ref. Bps75	9.50
Labour, say 20 min @ £11.76	3.92
Required rate	13.42

The door item was priced *pro rata* to item 96C, which was for a similar type of door but size 762 × 1981 mm. The rate was arrived at as follows:

		£
Bill rate, item 96C		61.87
Additional prime cost of 838 mm door	3.09	
Additional labour allowance	0.72	
	3.81	
15% OHP	0.57	
		4.38
Required rate		66.25

The door frame item was priced *pro rata* to bill item 95D which was for a door frame size 762 × 1981 mm, taking into account also other items for frames. The items in the bills were:

	£
50 × 88 mm Frame for door size 762 × 1981 mm	66.00
50 × 88 mm Frame for door size 838 × 1981 mm	67.85
63 × 88 mm Frame for door size 762 × 1981 mm	80.00

On a size proportion basis, the rate was determined at £82.24. The remaining items were priced at bill rates.

AI No. 48 Omission of bookcase units

The omissions formed a complete section of the original taking-off and cols. 325–336 of the dimensions were marked 'X' at the beginning and end of the section for ease of reference. The items involved formed a separate group in the Furniture/Equipment section of the bills and were omitted in the variation accounts as a group together with the related decoration items.

The consequential additions items were all deduction items in the original dimensions, and, upon identification for omissions purposes, they were converted into additions items with positive quantities (see comments under AI No. 23, p. 150).

References

1. *Standard Method of Measurement of Building Works, Seventh Edition Revised 1998* (London: RICS Books, 1998), p. 11, rule 1.2.
2. *Definition of Prime Cost of Daywork carried out under a Building Contract* (London: RICS Books, 1975).
3. NDEKUGRI, I.E. and RYCROFT, M.E., *The JCT 98 Building Contract: Law and Administration* (Butterworth-Heinemann, 2000), p. 138.
4. RICS, Questions and Answers, *Chartered Quantity Surveyor*, Vol. 8, No. 8, 1986, p. 8.
5. SIMS, J., Contracts, *Building*, Vol. 228, No. 6, 1975, p. 83; Vol. 229, No. 49, p. 72.
6. JCT, *Consolidated Main Contract Formula Rules* (London: RIBA Publications, 1987).
7. Building Cost Information Service, *Schedule of Basic Plant Charges* (Kingston-upon-Thames: RICS, 2001).
8. KEATING, D., *Law and Practice of Building Contracts* (London: Sweet & Maxwell, 1978), p. 315; WALKER-SMITH, D., *The Standard forms of Building Contract* (London: Charles Knight & Co. Ltd, 1971), p. 43.

8

Fluctuations

Introduction

Any contract, other than a prime cost contract, will be either 'firm price' or 'fluctuating price'. *Firm price contracts* are those where reimbursement of changes in costs is limited to those relating to statutory contributions, levies and taxes. *Fluctuating price contracts* are those containing provisions for reimbursement of changes in a wide range of labour costs and materials prices in addition to statutory costs. It should be recognized that the term 'fluctuations' includes both increases and decreases and that they may lead to payments becoming due to the contractor from the Employer and vice versa.

The Joint Contracts Tribunal (JCT) Form contains three different provisions, any one of which may be incorporated in a contract by deleting the other two clause references in the Contract Particulars. These provisions are

(a) adjustment of the contract sum limited to fluctuations in statutory contributions, levies and taxes, as specified in clause 4.21, Fluctuations Option A;
(b) adjustment of the contract sum in respect of fluctuations in labour and materials costs and in statutory contributions, levies and taxes, as specified in clause 4.21, Fluctuations Option B; and
(c) adjustment of the contract sum in respect of fluctuations as in (b) above by the use of price adjustment formulae, as specified in clause 4.21, Fluctuations Option C.

(a) and (b) above relate to what may be called the *traditional method* of reimbursement of the actual amount of fluctuations allowed under the respective clauses of the contract, and (c) to the *formula method* of calculating, by use of formulae, a sum to compensate for loss due to fluctuations.

Traditional method

The object of the traditional method is to ascertain the actual amount of the fluctuations in costs and prices within the scope of the appropriate clauses of the contract, as given above. That amount will often be less, sometimes much less, than the total amount by which costs and prices have really fluctuated, due to the limitations imposed by the operation of the contract clauses. Those costs and prices may be considered under the three main headings referred to in the JCT Form.

Statutory contributions, levies and taxes relative to 'workpeople' and others

Fluctuation Options A and B are similar and provide for the reimbursement of fluctuations occurring after the Base Date in (i) the amounts of any statutory contribution, levy or tax which, at the Base Date, the contractor is liable to pay or (ii) the amount of refund of any such contribution, levy or tax which is receivable by the contractor at the Base Date, as an employer of workpeople (as defined in Options A and B). The Base Date is defined as the date stated in the Contract Particulars (see clause 1.1).

An example of these costs is employers' National Insurance contributions. The wording of the clauses provides not only for changes in contributions, etc., payable at the Base Date but also for new types of contribution or tax which may be imposed and for existing ones which may be abolished.

The provisions of Option A relate not only to operatives employed on site but also to

(a) persons who are directly employed elsewhere by the contractor in the production of materials or goods for the Works, such as joinery shop workers and
(b) persons, such as timekeepers, site agents, contracts managers and site surveyors, who, although employed by the contractor

on site, are not 'work-people' as defined in the clause. Amounts in respect of this class of persons are the same as for a craftsman, but in proportion to the amount of time spent on site per week, within certain specified limitations.

Labour costs in addition to statutory costs

Option B.1 deals with fluctuations in labour costs other than those imposed by statute. In this sub-clause, the basis of the prices in the bills or schedule of rates is first defined and then it is provided that any increases or decreases within the prescribed limits shall be paid to or allowed by the contractor.

The basis of adjustment is 'the rules or decisions of the Construction Industry Joint Council or other wage-fixing body applicable to the Works and which have been promulgated at the Base Date'. It is important to note the use of the word 'promulgated' in the preceding sentence. This means that if future wage rates have been agreed and published by a wage-fixing bodies, then the contractor is deemed to have made an allowance for these future increases in his rates. There have been instances where wage settlements have been agreed spanning a two- or three-year period. The CIJC rules and conditions are published in their Working Rule Agreement for the Construction Industry. This agreement sets out basic wage rates, extra payments to which operatives may be entitled and various matters relating to working conditions. It also includes the provisions of the Annual Holidays Agreement and the amounts of weekly holiday credits for each operative for which an employer is liable. Specific reference is made in Option B.1.1 and B.1.2 to employer's liability insurance and third party insurance, which brings the respective premiums within the reimbursement provisions, for the reason that the premiums are usually calculated as a percentage of a contractor's wages bill.

It should be noted that the rates of wages referred to are those in accordance with the rules or decisions of a recognized wage-fixing body. Where operatives are paid at higher wage rates than the nationally negotiated rates, the basis of the prices in the contract bills or schedule of rates is no longer as stated in Option B.1.1 and therefore Option B.1.2 cannot follow.[1]

The contractor is entitled to recover fluctuations on transport charges in relation to workpeople engaged on the project as long as the contractor attaches a 'basic transport charges list' to the contract documents. Similarly the contractor is entitled to recover fluctuations on fares, which the contractor reimburses the workpeople, where the fares are reimbursable in accordance with the rules and decisions of a wage-fixing body.

Operatives employed by 'domestic' sub-contractors (whether on a labour-only basis or not) rank for recovery of fluctuations in the same way as, and on a similar basis to, the main contractor's directly employed labour,[2] provided that the contractor has complied with the requirement of Option B.4.1.

The classes of persons coming within the scope of Option B are same as those to which Option A applies, as described above.

Materials costs

Where Option B applies

The basis of adjustment is the market price of materials, electricity, fuels, etc., current at the Base Date. Fluctuations are allowable where there has been an increase or decrease in the market price of materials compared with the market price at the Base Date. Fluctuation in the market price of materials, etc., may come about as a result of changes in the general economy, changes in levels of competition, shortage of raw materials or changes in inflation which cause suppliers and manufacturers to adjust their prices. Market prices may also fluctuate as a result of a change in tax or duty on a material or from either the abolition of existing duty or tax or the imposition of new types of duty or tax. Fluctuations are allowable where there has been a change in duty or tax on the disposal of site waste or changes to import duties, purchase tax, etc., on materials, goods or fuels, etc. VAT is excluded from the fluctuation calculation where the contractor is able to treat it as an input tax, which means the contractor is able to recover the tax paid through his normal VAT returns.

Sometimes, claims are made in respect of changes in price, which are due to other causes than changes in market prices. These may be buying in part loads or small quantities, obtaining materials from another supplier at less favourable discounts or because the material is not the same as, although similar to, that in the contract documents. Adjustments to the contract sum should not be made for fluctuations in prices due to these or similar causes.

Option B.12.2 is specific that timber used in formwork ranks for fluctuations although it is not part of the permanent work, whereas other consumable stores such as plant and machinery may not be considered for fluctuations. The inclusion in a claim of materials drawn from the contractor's stores sometimes presents a problem, if the contractor is unable to produce invoices for them. Option B.3.2 is specific that the sum to be paid to or allowed by the contractor is to be the difference between the market price at the Base Date and the market prices current when the materials were bought, and which were payable by the contractor. Some surveyors will not include materials from stock for which the contractor cannot produce an invoice. Others will include them on assumed purchase prices equivalent to those being paid for other quantities of the same materials.

Materials used by 'domestic' sub-contractors rank equally with those used directly by the general contractor and on the same basis, provided that the contractor has complied with the requirement of Option B.4.1.

The surveyor should check the total quantities of each material listed in the claim to see that they are not in excess of those reasonably required, remembering to make allowances for reasonable losses due to wastage and theft. Any excess quantities should, of course, be excluded.

Where Option A applies

The basis of adjustment is the types and rates of duty and tax payable at Base Date, with reference to materials, goods, electricity, fuels, disposal of materials and disposal of site waste. Any adjustments allowable are limited to changes in types or rates of duty or tax relative to any of the above, and to price changes due to the abolition of a type of duty or tax or to the imposition of a new one. The comments made in the preceding section apply here also.

Additions to net sums recoverable

The sums calculated for fluctuations are not subject to any profit addition (see Option A.8 and B.9 of the JCT Form) but a percentage addition may be made added to the fluctuations in accordance with Option A.12 and B.13. The percentage addition, if any,

LIVERPOOL JOHN MOORES UNIVERSITY
LEARNING SERVICES

will be stated in the Contract Particulars. This addition applies to all fluctuation amounts recoverable under the provisions of Option A or B.

The percentage addition was introduced in order to go some way towards meeting the objections of contractors that only a small proportion of the true amounts of the increased costs actually incurred was being recovered.

Exclusions

The following kinds of work are excluded from the fluctuations provisions of Option A and B of the JCT Form.

(a) Work paid for on a daywork basis. This is because the rates used for valuing daywork should be rates current at the time the work is done. Those rates are used (and not tender date rates) because of the percentage additions made to the prime costs (see p. 125);
(b) Changes in value added tax charged by the contractor to the Employer; and
(c) Work paid for under a Schedule 2 Quotation (see clause 4.22).

Recording and checking fluctuations claims

It is the contractor's obligation to notify the architect of any fluctuations. The notification is to be given in writing within a reasonable time from when the fluctuation occurred. If the contractor wishes to recover a fluctuation then he must provide this written notification (see Option A.4.2 and B.5.2). It is also the contractor's responsibility to provide, at the architect's or quantity surveyor's request, evidence and calculations in support of a fluctuations notification (see Option A.7 and B.8).

It is in the interests of the contractor and the surveyor that a satisfactory system of recording and submitting the details of fluctuations claims be agreed at the start of the contract. This means that the system employed should facilitate the checking of the details from original invoices, wages sheets, etc. The amount of checking required can be considerable and the more systematic the contractor is in recording the data and preparing his claim, the sooner the checking and computation of the sums payable can be completed.

In order that amounts due may be paid without delay, the contractor should present claims at regular intervals, say, weekly or fortnightly. All supporting documents should be provided for the surveyor's inspection when checking the claim and it is useful to have the signature of the clerk of works, if there is one, on time sheets to certify the correctness of the names and hours shown. Alternatively, if a standard claim form is used to record the data taken from time sheets, provision may be made on the form for certification of the correctness of that data.

Figure 8.1 shows a summary of part of the materials claim relating to the contract of which particulars are given in Appendix A. The claim was prepared in accordance with the Contract, i.e., under the provisions of Option B. Only the increases in the cost of cement are shown, being part only of the complete materials fluctuations claim. The higher price of the November delivery was due to the small quantity and not to an increase in market price. It was therefore reduced by the surveyor to the rate appropriate to full loads.

Payment

As long as the contractor provided the necessary written notification, then payment of amounts recoverable is required to be made in the next interim certificate after ascertaining, from time to time, the amounts due. Such amounts are not subject to retention. The total amount of fluctuations will also be included in the final account as an addition or deduction, as the case may be.

Fluctuations after expiry of the contract period

A frequent cause for dispute in the past was whether the contractor was entitled to be paid for increases in costs relative to the period between the 'Date for Completion' and the date of actual completion. The contractor is contractually obliged to finish the works by the completion date, i.e., the date stated in the Contract Particulars or a later date where the architect has granted an extension of time. It is now clearly established that fluctuations

Materials fluctuation claim

Material: Cement
Basic price: £107.79 per tonne

Sheet no.: 1

Supplier	Invoice no.	Date	Quantity	Invoice price	Decrease	Increase	Total decrease	Total increase
Grey Square Cement Co Ltd	DF897	16.05.06	20 tonnes	112.80		5.01		100.20
Grey Square Cement Co Ltd	LX1376	11.07.06	10 tonnes	112.80		5.01		50.01
Greenthorne Cement Co Ltd	G2376X	15.09.06	12 tonnes	112.80		5.01		60.12
Greenthorne Cement Co Ltd	K1556Z	05.10.06	10 tonnes	112.80		5.01		50.01
Greenthorne Cement Co Ltd	K2485W	02.11.06	2 tonnes	*112.80* ~~125.00~~		*5.01* ~~17.21~~		*10.02* ~~34.42~~
Grey Square Cement Co Ltd	PR884	18.01.07	6 tonnes	112.80		5.01		30.06
Grey Square Cement Co Ltd	DL1285	20.03.07	6 tonnes	118.50		10.71		16.71
						Net total carried to summary		317.13 ~~341.53~~

Figure 8.1 Example of materials fluctuation claim

amounts are recoverable until the date of actual completion, that is, including the period after the date by which the contract should have been completed. The decision of the Court of Appeal in the case of *Peak Construction (Liverpool) Limited v McKinney Foundations Limited (1970), 69 LGR 1*, settled the matter and although the contract in that case was not based on the JCT Form, the Joint Contracts Tribunal has accepted the judgement as applying to the JCT Form.[3]

The JCT Form, in Option A.9.1 and B.10.1, 'freezes' the amount of fluctuations, however, at the levels, which are operative at the Completion Date (as stated in the contract or as extended). Consequently, although fluctuations should be calculated up to the date of actual completion, such calculations should be made on the basis of the changes in costs and prices occurring up to, but not after, the Completion Date.

Other non-recoverable fluctuations

It has already been pointed out (see p. 157) that where a contractor pays operatives at rates above the nationally negotiated rates, the provisions of Option B no longer apply. That clause applies only to 'rates of wages and the other emoluments and expenses ... payable ... in accordance with the rules or decisions of the Construction Industry Joint Council or some other wage-fixing body ...'. Any fluctuations in such inflated rates of wages are not recoverable, even where the fluctuations are related to nationally agreed rates.

The non-productive hours of overtime (even when expressly authorized) should not be included in allowable fluctuations payments. Only the actual hours worked rank for reimbursement.

Also excluded is the time spent on making good defects, e.g., under the provisions of clauses 2.38, 3.18, 3.19. Such work is required to be done entirely at the contractor's expense.

Formula method

In contrast to the traditional method of reimbursement of fluctuations, the object of the formula method is to calculate sum or sums, which will compensate the parties to the contract for loss incurred due to increases and decreases in costs. No attempt is made to calculate the *actual* amount of loss involved. Consequently, the sums recoverable by the formula method differ from those recoverable under the traditional method and will usually be greater.[4]

The formula method is incorporated in the 'with quantities', 'without quantities' and 'with approximate quantities' variants of the JCT Form, by means of Fluctuations Option C. It may also be used for schedule of rates contracts.

The JCT Form refers in Fluctuations Option C.1.1.1 to a separate document called the *Formula Rules*,[5] which sets out the formulae and defines their use. By virtue of the provisions of Option C, the *Formula Rules* are made part of the Standard Form and, therefore, contractually binding on the parties.[6]

The application of the formulae is by means of indices, compiled by the Construction Directorate of the Department of Trade and Industry and published on a monthly basis.[7] The indices measure the fluctuations in costs of building work classified in 60 categories. As an alternative to using individual categories, it is possible to combine two or more categories to create a work group. Since April 2001, a 'simplified formula' has been available that enables any number of weighted categories to be combined for the purposes of the fluctuation calculation. The procedures for using the simplified formula are explained and demonstrated in the Monthly Bulletin of Indices.[8] Figure 8.2 gives the list of Series 3 work categories. The indices are intended to be applied monthly to the sums included in interim valuations, thus ensuring that the payment of fluctuations is made promptly. The index number for each work category is based upon price movements of a number of selected items of labour, materials and plant, which are calculated by obtaining cost information from a variety of sources.

It follows, therefore, that the sums produced by the application of the indices to the prices in interim valuations must necessarily reflect a *general* fluctuation in prices and cannot relate to *actual* fluctuations in labour, materials and plant costs applicable to a particular contract.

Preparatory work necessary before using the formulae

The *Formula Rules* give a formula to be applied to each work category (Rule 9), a second formula has been prepared for use with

Series 3 Building formula
Work categories

01 Demolitions
02 Excavation & disposal
03 Filling: Imported, hardcore & granular
04 Piling: Concrete
05 Piling: Steel
06 Concrete: In situ
07 Concrete: Formwork
08 Concrete: Reinforcement
09 Concrete: Precast
10 Brickwork & blockwork
11 Stonework
12 Softwood carcassing & structural members
13 Metal: Decking
14 Metal: Miscellaneous
15 Cladding & covering: Glazed
16 Cladding & covering: Fibre cement
17 Cladding & covering: Coated steel
18 Cladding & covering: Plastics
19 Cladding & covering: Glass reinforced plastics
20 Cladding & covering: Slate & tile
21 Cladding & covering: Lead
22 Cladding & covering: Aluminium
23 Cladding & covering: Cooper
24 Cladding & covering: Zinc
25 Waterproofing: Asphalt
26 Waterproofing: Built-up felt roofing
27 Waterproofing: Liquid applied coatings
28 Boards, fittings & trims: Softwood
29 Boards, fittings & trims: Hardwood
30 Boards, fittings & trims: Manufactured
31 Linings & partitions: Plasterboard
32 Linings & partitions: Self-finished
33 Suspended ceilings
34 Raised access floors
35 Windows & doors: Softwood
36 Windows & doors: Hardwood
37 Windows & doors: Steel
38 Windows & doors: Aluminium
39 Windows & doors: Plastics
40 Ironmongery
41 Glazing
42 Finishes: Screeds
43 Finishes: Bitumen, resin & rubber latex flooring
44 Finishes: Plaster
45 Finishes: Rigid tiles & terrazzo work
46 Finishes: Flexible tiles & sheet coverings
47 Finishes: Carpets
48 Finishes: Painting & decorating
49 Sanitary appliances
50 Insulation
51 Pavings: Coated macadam & asphalt
52 Pavings: Slab & block
53 Site planting
54 Fencing
55 Pipes & accessories: Spun & cast iron
56 Pipes & accessories: Steel
57 Pipes & accessories: Plastics
58 Pipes & accessories: Copper
59 Pipes & accessories: Aluminium
60 Pipes & accessories: Clay & concrete

Figure 8.2 Price adjustment work categories

the *balance of adjustable work* (i.e., work ranking for adjustment but which is not capable of allocation to any work category, Rule 26) and a third formula for use for valuations after Practical Completion and in the Final Certificate (Rule 28). After giving corresponding formulae for use with work groups (Rules 29 and 38), five formulae applicable to five types of specialist engineering installations are given (Rules 50, 54, 58, 63 and 69). These formulae are for use where there are works of a specialist nature and the building formula would not truly reflect the nature of the works to be carried out and more importantly would not reflect the costs associated with this type of work. Formulae are provided for the following categories of specialist work

- Electrical installations.
- Heating and ventilating and sprinkler installations.
- Lift installations.
- Catering equipment.
- Structural steelwork.

In order to apply the appropriate formula to each work category, all the measured work items in the bills of quantities must be allocated to work categories. This should be done before tenders are invited, either by annotating the bills or by preparing schedules. Then, when interim valuations are done, the value of work included in them must be allocated accordingly. Annotating bills may be done by giving the work category number against each item in a column next to the item reference column.

There will be some work or goods, the value of which is included in a valuation, which cannot easily (if at all) be allocated to work categories or which is excluded from adjustment and this is dealt with as discussed below.

Balance of adjustable work

This is the balance of main contractor's work 'which is properly subject to formula adjustment but which is neither allocated to a Work Category nor is Specialist Engineering Work'.[9] Some of the items that would fall within this definition are preliminaries, water for the works, insurances. The amount of the adjustment to the balance of adjustable work for any valuation period is in the same proportion as the total amount of adjustment of work allocated to work categories bears to the total value of work in all work categories (see Form D in the following worked example).

Multiple category items

Sometimes a bill item, such as an alterations item in a 'Works on Site' bill, may relate to work falling within more than one work category. Such an item is allocated to the work category to which the largest part of the item relates.

Provisional sums

Provisional sums for work which will be paid at current prices (e.g., dayworks, work carried out by statutory undertakers and other public bodies) are not allocated to work categories but are separately identified as not subject to adjustment.

Provisional sums for work to be carried out by the general contractor are not allocated to work categories in the bills, but when the work has been executed, the measured items set against the provisional sum will then be allocated to appropriate categories or, if relating to specialist work, will be adjusted using the appropriate formula.

Fixing of items supplied by the Employer

'Fix only' items may be dealt with either by

(a) allocating the item to an appropriate work category;
(b) including the item in the balance of adjustable work; or
(c) applying an index specially created for the purpose.

Whichever method is to be used must have been previously specified in the bills or agreed with the contractor in writing.

Variations

All additions items priced at bill rates or analogous rates or at rates current at the Base Date will rank for adjustment and should be allocated to work categories. Variations valued on a daywork basis, as a Schedule 2 Quotation or otherwise valued at rates or prices current at the date of execution should be excluded from the operation of the formulae.

Unfixed materials and goods

The value of unfixed materials and goods are normally excluded from adjustment. Only when they have been incorporated into the work will they become adjustable in one of the ways described above. However, this rule does not apply in the specialist engineering formulae for, catering equipment, structural steelwork and lift installations. Finally, under Rule 27 b, it does state that if an early delivery of materials is required, it is possible to consider them for fluctuations where such special arrangements have been set out in the tender documents and recorded in the contract documents.

Claims for loss and/or expense

Claims for extra payments under clauses 4.23 and 3.24 are similarly excluded, as they will be valued at their full value.

Calculation of adjustment amounts

The appropriate formula is applied to each part of the value of work included in a valuation (after deducting the value of any work not subject to adjustment) when it has been apportioned into (a) work categories, (b) balance of adjustable work, and (c) specialist engineering installations. Such application is subject to the following considerations.

Provisional indices

Indices for any month when first published are provisional. Normally, firm indices are published three months later. The calculation of the adjustment amounts initially will also be provisional and will be subject to revision when the next valuation is done following publication of the firm indices.

Base month

The base month from which adjustment is calculated is the calendar month stated in the Contract Particulars, which normally is the one prior to that in which the tender is due for return. The published index numbers for the base month will be used in the application of the formulae.

Valuation period

The value of work carried out during the valuation period is substituted in the formula and this period is defined as commencing on the day after that on which the immediately preceding valuation was done and finishing on the date of the succeeding valuation.

The index numbers for the valuation period are those for the month in which the mid-point of the valuation period occurs. If the valuation period has an odd number of days, then it will be the middle day of the period; if it contains an even number of days, the mid-point will be the middle day of the period remaining after deducting the last day. For example, if valuation 12 took place on 14 September and valuation 13 takes place on 14 October – the valuation period will run from 15 September to 14 October, inclusive a period of 30 days. By deducing the last day, i.e., 14 October, the valuation period will be reduced to 29 days and the middle day of that period will fall on 29 September. Therefore September's indices will be used when calculating the fluctuations for valuation 13.

Non-adjustable element

Where a local authority is the employer, it is obliged to limit the amount of fluctuations allowed to the contractor under Option C, and this is achieved by stating a percentage non-adjustable element in the Contract Particulars. For example, if a non-adjustable element of 10% is stated that means the contractor will only be allowed to recover 90% of the fluctuations calculated. A side note in the Contract Particulars shows that it is the intention of the JCT that the non-adjustable element deduction should only apply where the employer is a local authority; it is not intended to be used for private employers.

Retention

The fluctuations amounts calculated by the formula method, unlike those ascertained by the traditional method, are subject to retention deductions. The reason for this is that what the formula method really does is to effect an updating of the rates and prices in the contract bills. The original rates and prices when included in valuations are subject to retention and it is logical, therefore, that the updated ones should be also.

Application of the formula rules to work executed after the Completion Date (or extended Completion Date)

Under the provisions of Option C. 6.1.1 of the JCT Form, the formula is applied to the value of work executed after the Completion Date (or extended completion date) but using the index numbers applicable to the month in which the Completion Date falls. This has the same effect as if all the work remaining to be done after the Completion Date had been carried out during the month in which the Completion Date occurred. It should be noted that this provision will not apply if either clause 2.26 or 2.29 has been amended or where the contractor has given a valid written notice of delay and the architect has failed to fix or confirm the Completion Date in accordance with the provisions of clause 2.28.

Adjustment of value of work included in interim valuations issued after the issue of the Certificate of Practical Completion

Rule 28 of the *Formula Rules* gives the formula to be used for adjusting the value of executed work which is included in interim certificates issued after the Certificate of Practical Completion has been issued, such value, in consequence, not being attributable to any valuation period. The formula effects an adjustment to such value in the same proportion as the total fluctuations included in previous certificates bears to the total value of work included in such certificates. In other words, the adjustment is calculated on the basis of the average fluctuations over the contract period.

This provision would apply to the difference between the actual value and an approximate valuation of variations and/or remeasured work and to the difference between the balance of adjustable work contained in the final account (and therefore in the Final Certificate) and that in respect of which fluctuations have already been included in interim certificates.

Errors

'Mechanical' errors

Possible errors, which may arise in the course of using the formula method, are listed in Rule 5a of the *Formula Rules*. They are

(a) arithmetical errors in the calculation of the adjustment;
(b) incorrect allocation of value to work categories or to work groups;
(c) incorrect allocation of work as Contractor's Specialist Work; and
(d) use of an incorrect index number(s).

If any such error has occurred, then the surveyor is required to correct it. Rule 5c requires that, when making the correction, the same index numbers be used as used when the original calculation was done, except that firm index numbers, if available, should be substituted for provisional ones.

Rule 5b provides for the contractor to be granted access, if he so requests, to the 'working documents' which show the calculations of the fluctuations included in an interim certificate for payment, so that he may satisfy himself that errors of the kind referred to above have not been made.

Errors in preparing a valuation

The *Formula Rules* make no provision for the retrospective correction of errors made in the course of preparing an interim valuation. Such 'errors' include the valuation of provisional items of work at the quantities given in the bills which upon remeasurement are found to be different.

The *Users Guide* advises that valuations should be as firm and accurate as possible, as these will not normally be reassessed at a later date.[10] The Quantity Surveyors (Practice and Management) Committee of the RICS has stated the view that, if errors have occurred in the preparation of a valuation, they should be corrected but that, once firm index numbers have been substituted for provisional ones, the facility for such correction will no longer exist.[11]

When errors of this kind are corrected in the manner envisaged by the RICS Quantity Surveyors Committee, the correction will be effected in the same way as described for 'mechanical' errors, that is, as set out in Rule 5c.

Any further corrections, which may remain to be made after firm index numbers have been applied, however, may be dealt with in the same way as the value of work included in interim certificates issued after Practical Completion (see above). In other words, adjustment of the errors may be made on the basis of the average of fluctuations over the whole of the contract period.

Examples

The following examples relate to the contract of which particulars are given in Appendix A. The application of the Formula Method, using work categories, is shown as it would have applied had the fluctuations provision been based on Option C. The Base Month would then have been stated in the Contract Particulars as July 2005. For the purposes of the example, Valuation No. 12 in the examples given at the end of Chapter 10 has been used. The indices used are the 1990 Series (Series 3).

Note: *Published indices for May to August 2006 were not available at the time this Chapter was being prepared, consequently a series of ad hoc indices have been created to enable the formula calculations to be demonstrated.*

It was necessary, prior to Valuation No. 1, to make preparatory calculations. The calculation proformas are based upon examples produced by the Department of the Environment[12] and require the following work to be executed

(a) to ascertain from the bills the value of work in each of the appropriate work categories and this was done on Form A (exclusions are preliminaries, provisional sums, builder's work

in connection with plumbing and builder's work in connection with electrical installation);
(b) to calculate the balance of adjustable work, as shown on Form B; and
(c) to enter on the form for the 'Calculation of Gross Fluctuations for Interim Valuations' the published Base Month index numbers for all the work categories involved.

The foregoing information was then available for use when each valuation was done.

When Valuation No. 12 was carried out on 14 September 2006, it would have been necessary to prepare a much more detailed valuation than the example provided in Chapter10. Where formula fluctuations are to be calculated, it is necessary to be able to identify all the relevant work categories within the valuation. This would mean for example, that the drainage work that was valued at £16,004 would have to be broken down into its various work categories (i.e., excavation and disposal; filling: hardcore, concrete, brickwork; accessories: cast iron and pipes: clay) and a value placed against each category. Therefore, it can be seen that where a project is operating under the formula fluctuations option, a surveyor will have to take much more care and time in preparing the interim valuations. However, time spent in preparing a detailed valuation will be amply rewarded when it comes to calculating the subsequent fluctuations. To make the fluctuations task even easier, it is possible to create the forms in a spreadsheet programme and prepare a number of templates containing the appropriate formulae and functions.

The following procedure was followed with reference to valuation 12.

(a) The gross value of the work executed to date, which could be allocated to work categories was entered in column 2 on the calculation form (see p. 179). Any extra work ordered through variations that has been priced at bill rates must also be analysed and allocated to the appropriate work categories (to simplify the example this step has been ignored, as has the need to adjust the value of work by the 2.395% tender adjustment). The gross value of work that had been completed previously was extracted from valuation 11 and the values entered in column 3. By deducting the values in column 3 from the values in column 2, it was possible to determine the value of work the contractor had executed for the valuation period. The resulting value was entered in column 4. In column 6, under the heading 'Provisional', the index numbers for August for the relevant categories were entered. In column 7 was entered the fluctuation allowance for each category using the expression

$$V \frac{I_v - I_o}{I_o}$$

where

V = value of work executed in the work category during the valuation period;

I_v = index number for the work category for the month during which the mid-point of the valuation period occurred;

I_o = the work category index number for the Base Month.

(b) The total of items assigned to work categories included in the valuation before adjustment and the amount of the adjustment were then entered at Q and R on Form D (see p. 180). The total value to date of the balance of adjustable work was calculated from the valuation and entered on Form D and from this the balance of adjustable work for Valuation No. 12 was calculated by deducting the amount included in Valuation No. 11. The gross adjustment on the balance of adjustable work was then calculated, as indicated on Form D, and the amount entered in the space beneath the amount in respect of the work categories total. The sum of these two amounts (£8,427.23 was the total of fluctuations for Valuation No. 12 using the provisional index numbers.

(c) At the same time as provisional index numbers for August became available, firm index numbers were published for May and so the fluctuations total included in Valuation No. 9, carried out on 14 June 2006, calculated on provisional index numbers, could then be adjusted. Accordingly, the calculation form, used to record the provisional fluctuations calculation for each work category, was then filled in under the column heading 'Firm' (see p. 181). The firm index numbers for May were entered in the 'Index for Valuation Period' column and, in those cases where they differed from the provisional numbers, the revised amount of 'gross valuation including VOP' was calculated. Where there was no difference, the figures were simply carried over from column 7 to 9. The revised amounts, as appropriate, were then entered on Form D (see p. 182) alongside the provisional ones and the difference in the totals at 'Y' and 'Z' (£95.14) was entered in the last space on the form as an addition for inclusion in Certificate No. 12.

It will be obvious from the foregoing, that it is essential that the record of calculations on the calculation form and Form D should be carefully filed away after each valuation, along with Forms A and B.

The calculations described above are shown in Figure 8.3.

Specialist engineering formulae

On this project there are a number of operations that come within the category of a specialist engineering formula, e.g., mechanical heating, security, electrical services and the lift installation. The lift installation would be subject to the 'Lift Installations' formula; electrical and security systems work would be subject to the 'Electrical Installations' formula and the mechanical heating would come under the 'Heating and Ventilating and Sprinkler Installations' formula. To illustrate the use of the specialist engineering formula it is proposed to look at the lift installation for this project.

Lift installations

For lift installations there are two formulae, one for shop fabrication work and another for the site installation work. The formula for the shop fabrication is applied either upon completion of the manufacture of the components or upon delivery to site, depending on what was agreed in the contract documents. The formula for the site installation is applied on completion of the installation on site. For this worked example, the shop fabrication work was completed a few months ago and the fluctuations have already been calculated for that element. The site installation work commenced

Form A

Allocation of bill of quantities items to work categories

Contract No. *265/6* | Employer *Skinton Development Co*

Project *Shops and Flats Thames Street, Skinton* | Contractor *Beecon Ltd*

Bill of quantities Nos.	2 – 4

Cat No. 3/	Bill of quantities item reference	Amount (£)
1	*16A–16F*	5,661
2	*20A–23G; 138A–138F; 142A–142B*	119,552
3	*24B–25C; 138G–139J; 142C–142F*	25,560
4	*26A–26G*	44,000
5		
6	*27A–28J*	122,868
7	*29A–30G*	19,147
8	*31A–31H*	40,390
9	*32A–32E*	6,470
10	*34A–43C; 130A–136E*	159,596
11		
12	*45A–68D*	71,574
13		
14		
15		
16		
17		
18		
19		
20	*70A–77C*	27,937
21	*78A–78G*	1,410
22		
23		
24		
25	*89A–89F*	50,062
26	*90D–91F*	15,548
27		
28	*113A–114C; 112–112D*	3,458
29		
30		
31		10,456
32		
	Total to overleaf (£)	723,689

Figure 8.3 Fluctuations calculations by formula method

Form A (cont.)

	Total from overleaf	723,689
33		
34		
35	*93A–102F*	124,238
36		
37		
38		
39		
40	*122A–123L*	2,990
41	*103A–104C*	4,002
42	*107A–107D*	6,213
43		
44	*105A–106P*	20,484
45		
46	*108A–108F*	3,824
47		
48	*109A–109S*	12,703
49	*110A–111K*	13,920
50	*113A–113L*	29,815
51		
52	*142N–143G*	8,466
53		
54	*141A–141F*	17,093
55		
56		
57	*117A–123C*	29,822
58	*125A–128E*	21,540
59		
60	*140G–140K*	12,950
	Total (£)	1,031,749
Provisional sums to which formula will apply		6,458
	Total L (£)	1,038,207
Provisional sums to which formula will not apply (including dayworks)	Total M (£)	44,230
Specialist engineering formula	Total N (£)	214,759

Figure 8.3 continued

Form B

(Part 1) **Summary of Form A**

Contract No. *265/6* Employer *Skinton Development Co*

Project *Shops and Flats*
Thames Street, Skinton Contractor *Beecon Ltd*

Bill of quantities No.	Work allocated to categories and provisional sums to which formula will apply (Total L Form A)	Provisional sums to which formula will not apply (including dayworks) (Total M Form A)	Specialist engineering formula (Total N Form A)
1			
2			
3	1,038,207	44,230	214,759
4			
5			
6			
7			
Totals (£)	X 1,038,207	Y 44,230	Z 214,759

(Part 2) **Calculation of balance of adjustable work**

		Contract sum (before deduction of credit for old materials)		£ 1,448,000
Deduct	(i)	Provisional sums to which formula will not apply (including dayworks) – Total Y above	£ 44,230	
	(ii)	Specialist engineering formula – Total Z above	£ 214,759	258,989
		Total of contract sum properly subject to price adjustment		£ 1,189,011
Deduct		Value of work allocated to work categories and provisional sums to which the formula will apply – Total X above		1,038,207
		Balance of adjustable work		£ 150,804

Figure 8.3 continued

Form WC1/WC2

Calculation of Gross Fluctuations for Interim Valuations

Project:	*Shops and Flats, Skinton*	Contract No:	*265/6*
Contractor:	*Beecon Ltd*	Employer:	*Skinton Dev. Co*
Valuation date:	14-Sep-06	Val No:	*12*
Previous valn. date:	12-Aug-06	Base month	Valn month
Mid point (days)	17	**July-05**	**Aug-06**
Valn mid point:	29-Aug-06		

	Value of Work Executed			**Work Category Index Series 3**				
					Provisional		**Firm**	
Work Category (1)	To date (2)	Previous Valuation (3)	This Valuation (4)	Base Index (5)	Valuation Index (6)	Prov Fluctuation (7)	Valuation Index (8)	Firm Fluctuation (9)
1	5,531	5,531	0	231	258	0.00		
2	108,672	104,789	3,883	224	250	450.71		
3	15,740	12,960	2,780	197	201	56.45		
4	44,000	44,000	0	199	214	0.00		
6	104,548	94,560	9,988	193	211	931.52		
7	3,452	2,580	872	199	214	65.73		
8	12,450	7,500	4,950	174	173	(28.45)		
9	5,077	2,050	3,027	168	181	234.23		
10	109,163	94,540	14,623	205	222	1,212.64		
12	42,139	21,580	20,559	173	182	1,069.54		
20	17,069	11,568	5,501	171	181	321.70		
21	1,200	600	600	168	177	32.14		
25	17,742	17,742	0	219	239	0.00		
26	3,972	1,580	2,392	195	206	134.93		
28	14,235	7,540	6,695	177	189	453.90		
31	9,226	7,850	1,376	212	228	103.85		
35	50,954	30,248	20,706	178	183	581.63		
40	1,125	450	675	174	175	3.88		
41	12,356	6,235	6,121	146	146	0.00		
42	3,110	1,540	1,570	206	220	106.70		
44	13,170	5520	7,650	235	255	651.06		
48	3,340	250	3,090	232	252	266.38		
49	2,914	350	2,564	177	186	130.37		
50	4,991	4,500	491	153	166	41.72		
52	804	350	454	189	202	31.23		
54	10,346	6,520	3,826	182	204	462.48		
55	540	120	420	176	186	23.86		
57	14,568	9,458	5,110	181	186	141.16		
58	8,418	4,458	3,960	166	185	453.25		
60	8,925	7,580	1,345	225	238	77.71		
		Total (Q)	135,228		Total (VT)	8,010.33	Total (VT)	

Figure 8.3 continued

Form D

Summary of price adjustment calculations	Valuation No.	12	Date of valuation: *14 September 2006*
			for month of: *August 2006*
	Contract No. *265/6*		
	Project: *Shops and Flats, Thames Street, Skinton*		

Valuation this month	£	135,228	(Q)

Balance of adjustable work

Total to date	£	89,383	
Total from previous valuation	£	82,345	
This valuation	£	7,038	(P)

		Price adjustment due based on	
		Provisional indices (£)	Firm indices (£)
Gross adjustment due (from WC2) (Total VT)	(R)	8,010.33	
Gross adjustment on balance of adjustable work (P/Q × R)		416.90	
	Sub total	8,427.23	
Deduct: Non-adjustable element (%)		–	–
Total provisional price adjustment due carried forward to Certificate No. 13	(Y) £	8,427.23	
Total firm price adjustment due		(Z)	
Balance of firm price adjustment due carried forward to Certificate No.		Add/Deduct*	

*Delete whichever is not applicable
If Y is > Z then Deduct
If Y is < Z then Add

Figure 8.3 continued

Form WC1/WC2

Calculation of Gross Fluctuations for Interim Valuations

Project: *Shops and Flats, Skinton* — Contract No: *265/6*

Contractor: *Beecon Ltd* — Employer: *Skinton Dev. Co*

Valuation date:	14-Jun-06	Val No:	*9*
Previous valn. date:	15-May-06	Base month	Valn month
Mid point (days)	15		
Valn mid point:	30-May-06	**July-05**	**May-06**

	Value of Work Executed			Work Category Index Series 3	Provisional		Firm	
Work Category (1)	To date (2)	Previous Valuation (3)	This Valuation (4)	Base Index (5)	Valuation Index (6)	Prov Fluctuation (7)	Valuation index (8)	Firm fluctuation (9)
1	5,531	5,531	0	231	237	0.00	238	0.00
2	98,745	98,745	0	224	229	0.00	230	0.00
3	109,250	109,250	0	197	198	0.00	196	0.00
4	44,000	44,000	0	199	202	0.00	198	0.00
6	74,550	58,970	15,580	193	200	565.08	200	565.08
7	1,250	950	300	199	201	3.02	201	3.02
8	10,100	6,800	3,300	174	173	(18.97)	171	(56.90)
9	4,354	1,455	2,899	168	171	51.77	171	51.77
10	89,450	75,400	14,050	205	211	411.22	211	411.22
12	20,450	15,350	5,100	173	176	88.44	176	88.44
20	17,769	11,568	6,201	171	173	72.53	174	108.79
21	500	350	150	168	175	6.25	175	6.25
25	20,714	12,390	8,324	219	228	342.08	230	418.10
26	1,000	1,000	0	195	199	0.00	198	0.00
28	5,200	3,500	1,700	177	177	0.00	177	0.00
31	5,900	3,450	2,450	212	215	34.67	214	23.11
35	28,400	19,430	8,970	178	182	201.57	182	201.57
40	100	100	0	174	173	0.00	173	0.00
41	5,800	3,230	2,570	146	144	(35.21)	144	(35.21)
42	1,000	1,000	0	206	207	0.00	207	0.00
44	3,500	2,400	1,100	235	237	9.36	237	9.36
50	4,000	1,300	2,700	153	155	35.29	157	70.59
54	6,520	4,850	1,670	182	182	0.00	181	(9.18)
55	100	100	0	176	180	0.00	180	0.00
57	8,450	5,125	3,325	181	186	91.85	186	91.85
58	3,900	2,400	1,500	166	183	153.61	183	153.61
60	6,995	3,200	3,795	225	225	0.00	225	0.00
		Total (Q)	85,684		Total (VT)	2,012.57	Total (VT)	2,010.48

Figure 8.3 continued

Form D

Summary of price adjustment calculations	Valuation No.	9	Date of valuation: 14 June 2006
			for month of: May 2006
	Contract No. 265/6		
	Project: Shops and Flats, Thames Street, Skinton		

Valuation this month	£	85,684	(Q)
Balance of adjustable work			
Total to date	£	71,402	
Total from previous valuation	£	65,394	
This valuation	£	6,008	(P)

		Price adjustment due based on	
		Provisional indices (£)	Firm indices (£)
Gross adjustment due (from WC2) (Total VT)	(R)	2,012.57	2,101.48
Gross adjustment on balance of adjustable work (P/Q × R)		141.12	147.35
	Sub-total	2,153.69	2,248.83
Deduct: Non-adjustable element (%)		–	–
Total provisional price adjustment due carried forward to Certificate No. 10	(Y) £	2,153.69	
Total firm price adjustment due	(Z)		2,248.83
Balance of firm price adjustment due carried forward to certificate No. 13	Add/~~Deduct*~~		95.14

*Delete whichever is not applicable
If Y is > Z then Deduct
If Y is < Z then Add

Figure 8.3 continued

in July 2006 and was completed in September 2006. The formula for the site installation is as follows:

$$C = V\left[\frac{0.23\left(\frac{L_{es} + L_{ec}}{2} - L_{eo}\right)}{L_{eo}}\right]$$

C = the amount of the formula adjustment calculated.
V = value of the lift installation in accordance with the formula rules.
L_{es} = index number for electrical labour for the month in which commencement of site installation occurred.
L_{ec} = index number for electrical labour for the month in which completion of the site installation occurred.
L_{eo} = index number for electrical labour for the Base Month.

Example

Electrical labour base index July 2005	2527
Electrical labour index for July 2006	2607
Electrical labour index for September 2006	2619

$$C = 71{,}235\left[\frac{0.23\left(\frac{2607 + 2619}{2} - 2527\right)}{2527}\right] = 557.59$$

References

1. The Query Sheet, *Chartered Surveyor, BQS Quarterly*, Vol. 2, No. 4, Summer 1975, p. 56.
2. Questions and Answers, *Chartered Surveyor, BQS Quarterly*, Vol. 4, No. 1, Autumn 1976, p. 15.
3. The Query Sheet, *Chartered Surveyor, BQS Quarterly*, Vol. 2, No. 2, Winter 1974, p. 27.
4. GOODACRE, P. E., *Formula Method of Price Adjustment for Building Contracts* (Reading: College of Estate Management, 1978), p. 81.

5. JCT, *Consolidated Main Contract Formula Rules* (London: RIBA Publications, 1987).
6. JCT, *Practice Note 17, Fluctuations* (London: RIBA Publications, 1982), Para. 8.
7. *Price Adjustment Formulae for Construction Contracts, Monthly Bulletin of Indices* (Tudorseed Construction Ltd).
8. Ibid
9. JCT, *Consolidated Main Contract Formula Rules,* op. cit.
10. *Formulae for Construction Contracts: Users Guide, 1990 Series of Indices* (London: HMSO, 1995).
11. *Price Adjustment Formulae for Building Contracts* (London: RICS Books, 1975), p. 8.
12. *Price Adjustment Formulae for Construction Contracts*, op. cit.

9

Claims

Contractor's claims may be of three kinds

1. Common law.
2. *Ex gratia*.
3. Contractual.

Common law claims

These arise from causes which are outside the express terms of a contract. They relate to breaches by the Employer or his agents of either implied or express terms of the contract, e.g., if the Employer in some way hindered progress of the Works or if the architect were negligent in carrying out his duties, resulting in loss to the contractor.

Ex gratia claims

These have no legal basis but are claims, which the contractor considers the Employer has a moral duty to meet, e.g., if he has seriously underpriced an item whose quantity has been increased substantially because of the variation which will in consequence cause him considerable loss. The Employer is under no obligation to meet such 'hardship claims' but may be prepared to do so on grounds of natural justice or to help the contractor where otherwise he might be forced into liquidation.

Contractual claims

These arise from express terms of a contract and form by far the most frequent kind of claim. They may relate to any or all of the following:

(a) fluctuations;
(b) variations;
(c) extensions of time; and
(d) loss and/or expense due to matters affecting regular progress of the works.

Fluctuations claims

These relate to increases in the costs of labour, materials and plant and to levies, contributions and taxes, which the contract provides for the contractor to be reimbursed by the traditional method in Options A and B of JCT 05. This has been fully explained in Chapter 8 and requires no further treatment here.

Claims arising from variations

These may relate to one or more of the following clauses:

(i) Clause 5.6.1.2 – the surveyor may have priced at bill rates variation items which are apparently similar to bill items but which were not executed under similar conditions to those envisaged at the time of tendering. For example, an Architect's Instruction changing the kind of facing bricks may affect the time taken for laying the bricks, and may lead to more sorting if there is a higher proportion of misshapen bricks or if they are more susceptible to damage in handling. The contractor would be entitled to an increase over the bill rates.
(ii) Clause 5.6.3.3 – the contractor may claim that adjustment of one or more Preliminary items should be made, though the surveyor has not included the adjustment in his valuation of a variation. For example, an Architect's Instruction changing the pointing of faced brickwork from 'pointing as the work proceeds' to 'pointing on completion' may require scaffolding to be left standing for a longer period than would otherwise have

been necessary. If scaffolding is priced in the Preliminaries (as is usual), the relevant item should be adjusted.

(iii) Clause 5.9 – it may be claimed that, in consequence of a variation, the conditions under which other work is carried out has changed and therefore the bill rates for that work should be adjusted to reflect the changed conditions. Thus, a variation changing the part of the foundations of a building from concrete deep strip to short-bored piles would, it might be claimed, so reduce the total volume of concrete required for the rest of the deep strip foundations that the bill rates for the concrete would no longer apply. If this argument can be substantiated, the rates should be adjusted.

(iv) Clause 5.10 – the contractor may claim that he has incurred additional costs which are directly associated with a variation but for which he has not been reimbursed by the application of the normal valuation rules. An example of such a claim is given on p. 130.

Claims for extensions of time

These arise from clauses 2.26 to 2.29 and 5.3 of JCT 05. Clause 2.4 requires the contractor to complete the Works on or before the Date for Completion stated in the Contract Particulars (see p. 394), or such, later date as may be fixed by the architect under clauses 5.3 or 2.28.1. If the contractor fails to do so, he becomes liable for liquidated damages (clause 2.32), which the Employer is entitled to deduct from payments due to the contractor at the rate stated in the Contract Particulars for the period between the date when completion should have taken place and when it actually took place. This action of the Employer is subjected to the prior issue by the architect of a certificate of non-completion by the Completion Date and a written notice from the employer informing the contractor that liquidated damages may be withheld.

The rate of liquidated damages stated in the Contract Particulars should be a genuine pre-estimate of the likely loss to the Employer due to the time overrun and is not adjustable according to the actual loss incurred. The purpose of granting extension(s) of time is only to relieve the contractor of liability to pay liquidated damages for the period of the extension and does not carry an automatic right to reimbursement of any loss or expense, which the contractor may claim he has suffered or incurred due to the matter for which the extension was granted.

Clause 2.29 lists Relevant Events, the happening of any one of which is a ground for extending the contract time by such period(s) of time as the architect estimates is 'fair and reasonable', if it is 'apparent that the progress of the Works is being or is likely to be delayed'. The contractor must give prompt written notice of such actual or likely delay, (i) stating the cause(s), (ii) identifying any Relevant Events, (iii) giving particulars of the expected effects, and (iv) stating the estimated extent of the delay in the completion of the Works. The architect is under no obligation to act until such notice has been received.

Having received a notice the architect should endeavour to notify the contractor as soon as is reasonably practicable whether or not an extension of time is to be granted; but at the very latest must inform the contractor of his decision within 12 weeks of receiving the required particulars. If the period of time, from when the architect received the required particulars up to the completion date, is less than 12 weeks the architect should try and notify the contractor before the completion date. The architect must give a fair and reasonable extension of time if he is of the opinion that the cause of delay is a Relevant Event and that the completion of the Works is likely to be delayed as a result. Having once granted an extension of time and having set a new completion date, the architect is entitled on subsequent occasions to take into account any instruction for the omission of work issued after the fixing of that new completion date. For example a contractor provides evidence to show that he will suffer a three-week delay because of an instruction changing the roof design. However, in the period since the architect granted the previous extension of time he had issued an instruction omitting a section of external works that would reduce the contract period by one week. Therefore, the extension of time the architect grants on this occasion should reflect the omitted work. The architect should grant an extension of two weeks, and he should notify the contractor of the relevant omission used in the calculation. It is important to be aware that an architect, under clause 2.28, cannot set a revised completion date earlier than the completion date stated in the contract particulars regardless of the omissions that have been instructed (clause 2.28.6.4).

After the Completion Date the architect must review his previous decisions (if any) on extensions, and he must then notify the contractor in writing that he has

(i) fixed a later Completion Date if he considers that to be fair and reasonable in the light of any relevant events that have occurred,
(ii) fixed an earlier Completion Date in response to 'Relevant Omissions' that have been instructed since the last occasion an extension of time was granted, or
(iii) confirmed the Completion Date previously fixed.

Where the Completion Date occurs before the contractor has achieved practical completion the architect may provide the above notification at any time, from the Completion Date until 12 weeks after the date of practical completion.

Sectional completion

The above procedures relating to the granting of an extension of time will equally apply to the contract works where the sectional completion option is completed in the Contract Particulars resulting in a number of 'Dates for Completion of Sections'. In this instance, the contractor and architect will have to view the impact that a relevant event may have on each individual section and its relevant completion date.

Schedule 2 Quotation

When the contractor provides a Schedule 2 Quotation, he must also make due allowance for any adjustment to time required for carrying out the Works. This would normally take the form of requesting an extension to the contract period, but it is also possible for a contractor to indicate a shorter contract period is required. Through this procedure the contractor is able to have the completion date changed to one earlier than that stated in the Contract Particulars. If the architect accepts the Schedule 2 Quotation, he must confirm the adjustment of time required by the contractor and set a new completion date.

Claims for loss and/or expense due to matters affecting regular progress

These claims arise under clauses 4.23 to 4.26, 5.3 and clause 3.24 of JCT 05. Clause 4.24 lists five key events (in addition to deferment of giving possession of the site to the contractor under clause 2.5) which might materially affect progress. If the contractor believes that any one or more of those matters has or may affect the regular progress of the Works and that he has or may, in consequence, incur loss and/or expense, then he may apply in writing to the architect for reimbursement. If the architect is of the opinion that such loss and expense has been or is likely to be incurred due to one or more of the stated matters, he must ascertain or (as normally happens) he must ask the quantity surveyor to ascertain the amount of such loss and expense which has been or is being incurred.

All the matters listed in clause 4.24 are those for which the Employer is responsible, being acts or omissions of the Employer or the architect as his agent. Excluded from recovery are all other causes of loss or expense, such as those which are unknown or undefined or the result of commercial risks, bad pricing of the contractor's tender, bad weather or bad organization or management of the contract. Nor is there any automatic right of recovery of loss solely because an extension of time has been granted. On the other hand, it is possible for the contractor to have suffered recoverable loss or expense when no extension of time has been granted.

The architect need take no action unless and until a written application which clearly states the circumstances which have caused the loss has been received from the contractor. A telephone request or an oral application made to the architect on the site is insufficient; also a new application is necessary for each new matter that arises. If the architect fails to act upon receipt of an application properly made, the Employer will become in breach of contract and liable for damages.

The contractor is not required to formulate a claim, but must provide all necessary information to enable the amount of the loss or expense to be ascertained. This means that the contractor must be prepared to reveal what he probably considers to be confidential information about such matters as actual wages and bonuses paid to operatives, details of head office overheads, etc. 'Ascertain' means 'to find out exactly', so it is necessary for the quantity surveyor to find out the actual loss and not to rely upon estimates or the use of formulae in order to arrive at a sum. However, this strict interpretation must be viewed within a commercial context; a surveyor should satisfy himself as to the reasonableness and adequacy of a contractor's claim. In certain circumstances it may be necessary for an Employer to accept a formulaic approach or accept claims without a detailed breakdown (e.g., see sections on Head office overheads and Global claims).

The loss and/or expense, to be recoverable, must have been a direct consequence of the matter(s) referred to in the claim without there having been any intervening cause. Thus, if the Employer fails to provide materials for the Works which he has agreed to provide and the contractor is asked to obtain them, then the contractor is entitled to recover the expense incurred in so doing, as well as the loss due to delay in obtaining them. If, however, in obtaining those supplies, deliveries are held up because of a strike in the haulage industry and additional loss or expense ensues, this would not be recoverable, being an indirect result of the original cause.

If the contractor does formulate a claim or even if he just provides required information, he is much more likely to succeed in his claim if he makes a clear and orderly presentation of the data. It will not help the architect in forming an opinion as to the soundness of a claim or the surveyor in ascertaining the proper amount if they are presented with a jumbled mass of papers, which it is virtually impossible to sort out.

Heads of claim may be any or all of the following:

(i) insufficient use of labour and/or plant;
(ii) increases in cost of labour, materials etc. during the period of disruption;
(iii) site running costs;
(iv) head office overheads;
(v) finance charges and interest;
(vi) loss of profit.

(i) Inefficient use of labour and/or plant – it refers to men and plant standing idle or working at a reduced level of output. It would be necessary for the contractor in substantiating a claim to produce evidence of the estimated levels of output used in preparing his tender and records of the actual output during the disruption period.

(ii) Increases in costs of labour, materials, etc. during the period of disruption – increases in costs would normally only apply in the case of firm price contracts, i.e., Fluctuations Option A.
(iii) Site running costs – these relate to site staff costs, offices, mess-rooms, sheds, rates on temporary buildings, etc., normally referred to as Preliminaries. Prolongation of a contract could result in some or all of these items being required for a period in excess of that for which the contractor allowed in his tender. It is important to check that some of these items have not already been allowed for when pricing variations.
(iv) Head office overheads – these include the cost of maintaining head and branch offices, plant and materials yards, rents, rates, directors' and staff salaries, office running expenses, travelling expenses, professional fees and depreciation. During the normal course of a business these overheads are normally recouped by calculating their anticipated cost over the financial year and identifying an average percentage to be applied to estimates which would allow the overheads to be recovered. As a result of disruption occurring on a project, a contractor may consider he is obtaining an inadequate return on his head office overheads. This may arise, for at least two reasons:
   1 Because of delay and disruption the contractor is prevented from taking on other projects, thereby reducing his turnover and subsequently his recovery on overheads. This is sometimes referred to as unabsorbed overheads.[1] It may be argued that the contractor should increase his percentage mark up to recover his costs on a reduced turnover. The counter argument is that this would make the contractor less competitive and depress his turnover even further.
   2 Because of delay and disruption the contractor may have had to increase his management levels beyond what was originally anticipated, possibly as a result of having to monitor and administer the claim. This is sometimes referred to as dedicated overheads.[2]

   For many years surveyors working for a client would dismiss head office overheads as part of a loss and expense claim on the basis that the costs were too far removed from the cause of the claim and were not therefore a direct loss and/or expense. This argument was eventually proved to be incorrect through the development of case law (i.e., *Wraight Ltd v PHT*

(Holdings) Ltd (1968) 13 BLR 26 and Peak Construction (Liverpool) Ltd v McKinney Foundations Ltd (1970) 1 BLR 114). Therefore unless the contract conditions specifically exclude the recovery of unabsorbed overheads they should be accepted as a valid claim.

Because of the complexity of trying to sort out overhead costs, a formula is often used as a means of evaluation. Three in common use are Hudson's, Emden's and the Eichleay formula.[3] The first two are similar

$$\frac{h}{100} \times \frac{c}{cp} \times pd$$

where h = head office overheads and profit per cent included in the contract (Hudson's), or h = per cent arrived at by dividing total overhead costs and profit of the contractor's organization as a whole by total turnover (Emden's); c = contract sum; cp = contract period in weeks; and pd = period of delay in weeks.

Many surveyors do not accept the use of formula on the grounds that actual cost only is admissible and that the contractor must specify precisely what actual additional expense has been incurred.[4] The decision in *Tate & Lyle Food Distribution Co Ltd v Greater London Council (1982) 1 WLR 149* supports this view.

However, in the case of *J E Finnegan Ltd v Sheffield City Council (1988) 43 BLR 124*, it was held that the Hudson formula should be used to calculate head office overheads and profit as part of the plaintiff's claim. However, the learned judge, in referring to it, was confusing the Hudson formula with the Emden formula.

A summary of the situation is that losses attributable to head office overheads is a valid head of claim. There are two ways by which the losses may be quantified, by formula or by an assessment of actual costs. A contractor would be advised to assess actual costs wherever possible, but where this is not practicable the parties should be prepared to accept and negotiate on a formulaic approach. It should be remembered that a contractor does not have an automatic right to claim for loss of head office overheads. A contractor must be able to produce reasonable evidence that other work was available for tender, that there was a reasonable chance of successfully tendering for some of the work, but it was unable to tender because of resources being tied up on the project for which the claim is being submitted. Failure to provide this proof should result in this head of claim being dismissed (*Amec Building Ltd v Cadmus Investment Co (1996) 51 ConLR 105 and City Axis Ltd v Daniel P Jackson (1998) 64 ConLR 84*).

(v) Finance charges or interest – as with the previous head of claim the inclusion of interest on a loss and expense claim used to be a contentious item and was almost invariably dismissed by a client's surveyor. Again case law has demonstrated that it is in fact a valid claim (*F G Minter Ltd v WHTSO (1980) 13 BLR 1 and Rees and Kirby Ltd v Swansea BC (1985) 30 BLR 1*). Finance charges are interest charges incurred by the contractor on money he has had to borrow (or interest that he was prevented from earning on his own capital) in order to finance the direct loss and/or expense claimed. The rates of such interest must be those actually paid to the bank or other finance source and certified by the contractor's auditors. Interest due to the prolongation of payment of sums owing (such as retention monies) is not admissible.

(vi) Loss of profit – this refers to profit, which the contractor could have earned but was prevented from earning as a direct result of one or more of the matters listed in clause 4.24. For example, because of the prolongation of one contract, the contractor was prevented from taking on another. If he could prove his loss of profit as arising directly from a clause 4.24 matter, he would be entitled to reimbursement. This is very similar to a contractor's claim for unabsorbed overheads and the two items are often dealt with together.

It should be noted that acceleration costs, i.e., additional costs incurred due to speeding up progress of a contract in order to meet the Completion Date, are not recoverable. However, if an architect were to request a Schedule 2 Quotation for the rescheduling of the works the contractor could price the works on the basis of an earlier completion date (see Schedule 2, Section 2.2) with the result that the works are accelerated and the contractor is reimbursed. A contractor is not normally entitled to claim the costs of preparing a claim or providing detailed information to enable loss or expense to be ascertained.[5]

Global claims

When a contractor is presenting information in support of a claim it will be necessary for him to identify the contract conditions and events that have caused the claim, along with details of the effect the event(s) has had on the project and contractor. Unfortunately because of the complex nature of construction work it can be very difficult for a contractor to provide a breakdown of the loss and expense incurred against each separate event that has caused the disruption, i.e., identifying cause and effect. As a consequence a contractor may review the project as a whole, identifying a number of events that entitle a loss and expense payment to be claimed and then provide an overall costing for all these events, i.e., a global claim. Purists will argue that a global claim is not admissible because costs have to be clearly associated and identified against each individual event, an argument that was supported in *Wharf Properties Ltd and Another v Eric Cumine Associates and Others (1991) 52 BLR 1.* However, the general consensus appears to be that a grouping together of events into a global claim can be acceptable where it is impracticable to separate the events and costs, *Mid-Glamorgan County Council v J Devonald Williams and Partner (1991) 29 ConLR 24.* Therefore, a contractor would be best advised to try and associate costs with individual events as much as possible and then sweep up the remaining items that cannot be easily separated into a global claim.

Payment

Having ascertained 'from time to time' amounts of loss and expense incurred, the architect is required to include them in the next interim certificate after each ascertainment. Such amounts should be paid in full, i.e., they are not subject to retention. They will also be included in the final account for adjustment of the contract sum.

Schedule 2 Quotation

It is not normally acceptable to make any allowance for direct loss and/or expense when valuing a variation (see clause 5.10.2). An exception to this rule is where the contractor provides a Schedule 2 Quotation (see Schedule 2, item 2.3), in which case he is obliged to include a proper allowance for loss and/or expense associated with the variation. If the contractor fails to make a proper allowance for loss and/or expense in the Schedule 2 Quotation, he is not able to recover the shortfall by using clause 4.23. This is one of the risks the contractor accepts when providing a Schedule 2 Quotation.

Antiquities

The provisions of clause 3.24 in regard to loss and/or expense incurred as a result of disruption following the finding of antiquities, fossils, etc. are similar to the provisions in clause 4.23. The main difference is that the contractor is not required to give written notice to the architect as a pre-condition to consideration of a claim, although it would be advisable to inform the architect, so he may ascertain the amount of any direct loss and/or expense.

Bibliography

1. POWELL-SMITH, Vincent, *Problems in Construction Claims* (London: Blackwell Science, 1990).
2. TRICKEY, G., *Presentation and Settlement of Contractors' Claims, Second Edition* (London: Spon, 1996).
3. NEWMAN, P., *Loss and Expense Claims Explained* (London: RIBA Publications, 1994).
4. KNOWLES, R., *Claims: Their Mysteries Unravelled, Second Edition* (London: Knowles Publications, 1993).
5. POWELL-SMITH, Vincent, SIMS, John and CHAPPELL, David, *Building Contract Claims* (London: Blackwell Science, 1996).

References

1. Society of Construction Law, *The Society of Construction Law Delay and Disruption Protocol*, October 2002, www.scl.org.uk.
2. ibid.
3. KNOWLES, R., Calculating office overheads, *Chartered Quantity Surveyor*, Vol. 8, No. 5, 1985, p. 207.
4. MORLEDGE, R., Letter, *Chartered Quantity Surveyor*, Vol. 8, No. 7, 1986, p. 5.
5. RICS, *Contractor's Direct Loss and/or Expense* (London: RICS Books, 1987), p. 11.

10

Interim valuations

When the value of contracts was more than a few thousand pounds, it had always been normal practice for contractors to be paid sums on account as the construction work proceeded. This is because it was generally recognized that it would be unreasonable to expect contractors to finance construction operations without assistance from Employers. Also, the expense of borrowing large sums, which otherwise would be involved, would add significantly to total costs – which it is in the interest of Employers to minimize. Consequently, all the Standard Forms included provisions for periodic or 'interim' payments to be made for these reasons. However, as the result of legislation it is now a statutory requirement that most construction contracts make allowance for interim, stage or periodic payments.

The Housing Grants, Construction and Regeneration Act 1996 (HGCRA) came into force on 1 May 1998, and was claimed to be 'the most significant piece of legislation relating to the construction industry for decades'.[1] For the construction industry one of the most important areas of the Act was Part II, which referred to 'Construction Contracts'. The main features of the Act are that construction contracts entered into after 1 May 1998 must make adequate provision for payment and allow for disputes to be referred to adjudication. When the legislation was being prepared it was acknowledged that not all contracts would automatically incorporate the contract conditions that were required by the Act. For example, it is not uncommon for many domestic sub-contracts to be put in place with the minimum of paperwork, therefore to complement the Act the legislatory body also produced supporting regulations, 'The Scheme for Construction Contracts (England and Wales) Regulations 1998'. The purpose of the Scheme is to provide a set of regulations which would be applied to any construction contract that failed to properly incorporate the provisions of the Act. As a result it is not possible to avoid the Act by failing to incorporate the necessary provisions. Parties preparing contracts have to be aware that they must either incorporate the provisions required by the Act or be prepared to accept the regulations contained within the Scheme.

As a consequence the Act had a significant impact upon construction contracts in relation to payment procedures and most standard forms of contract were amended to ensure that they complied with the requirements of the Act. However, there may be instances where parties contract on a non-standard form or a heavily amended standard form, in which case it will be important that the parties are aware of the requirements of the Act to ensure that the payment procedures do in fact comply with the statutory requirements.

Payment requirements under the Act

There must be provision within an agreement that allows for periodic payments as the work progresses. The exceptions to this are where:

- the contract specifies that the duration of the work is to be less than 45 days, or
- the parties agree that the duration of the works is estimated to be less than 45 days.

The Act imposes a very flexible timetable on the payment provision; it initially states that the parties are free to agree:

- the intervals or circumstances when payments become due (there is no time limit imposed on how long the period may be between the date the amounts become due and the final date by which they must be paid);
- the amounts of the payments (but there must be a mechanism to determine how these amounts are to be calculated).

Within 5 days from the date a payment becomes due the payer must give the payee a written notice confirming the amount of money that he has already paid or that he intends to pay. In the notice he must give details explaining the basis on which the amount was calculated.

On occasions the party making the payment may wish to withhold monies, e.g., set-off. This is legally acceptable, but the Act requires that prior notice must be given. The notice must be given no later than the specified time period (as agreed between the parties) before the final date for payment. The notice must state the amount that it is proposed to withhold and the reason why the amount is to be withheld. If there are a number of reasons provided then the amount must be broken down to clearly show the sum of money allocated to each reason.

Suspension

If a payee does not receive full payment of the amount due, and there has been no effective notice given to state that monies were being withheld, then the payee has the right to suspend the performance of their obligations under the contract with the payer. Before this right of suspension may take effect the payer must be given a written notice advising them of the intention to suspend performance and informing them of the ground(s) for the suspension. The Act requires a minimum notice period of 7 days. If the payer subsequently pays the amount in full then the right to suspend performance is extinguished.

Conditional payment provisions

For many years sub-contract organizations have had to accept a term in their sub-contracts stipulating that a payment would only be made to them subsequent to the contractor receiving payment from the Employer. The Act has now outlawed this practice. It is no longer legally acceptable to state that a payment is conditional upon the payer receiving monies from a third party. The only exception to this rule is where the third party who is making the payment is insolvent, in which case a 'pay when paid' clause would be valid.

Failure to comply with the Act

If a contract does not make proper provision for the requirements of the Act then 'The Scheme for Construction Contracts (England

and Wales) Regulations 1998' will apply. With regard to the payment provisions, the Scheme will apply only to those parts where the contract has failed to meet the requirements of the Act. For example, the Scheme may be used to fill in a gap in the payment procedures or override a non-compliant term. Therefore, if a form of contract meets all the necessary payment provisions, with the exception that there is no provision for the 'giving a notice of intention to withhold payment', then Part 2, section 10 of the Scheme will be implied. The implied term will be that a notice of intention to withhold payment must be given at least 7 days before the final date for payment.

The Scheme

The following provide brief details of the Scheme in relation to payment procedures.

Interim payments

The amount of any periodic payment is calculated by determining the total value of work executed, plus the value of materials on site (only if this is allowed under the contract) and any other amount specified as being payable under the contract. All previous payments are then deducted and the difference is the amount due. However, the total amount of payments made under this procedure must not exceed the contract price.

Dates for payment

An interim payment will become due from the later of the two events:

- 7 days from the end of the 'relevant period' (i.e., a period specified in the contract conditions or if no period is defined – 28 days);
- the submission of a claim for payment by the contractor, i.e., a written application from the contractor specifying the amounts due and the basis of their calculation.

The final payment for a contract will become due from the later of the two events:

- 30 days from the completion of the work;
- the submission of a claim for payment by the contractor.

The Employer must pay any of the above amounts within 17 days from the date the payments became due.

Payment notices

Within 5 days, from the date a payment was due, the Employer must give the contractor a notice specifying the amount of the payment that has been made (or is to be made) and providing details of how the amount was calculated.

If the Employer intends to withhold monies from the amount due he must give the contractor prior notice. The notice is to be given not later than 7 days before the final date for payment.

Interim valuations under JCT SBC/Q

The JCT along with most other contract publishing organizations has fully complied with the payment requirements of the Act, and as a result the Scheme will not apply to its standard forms of contract.

Frequency of valuations and 'valuation date'

Clause 4.9 of the JCT Form provides for the Employer to pay the contractor such sums as are stated to be due in Interim Certificates issued by the architect at the periods stated in the Contract Particulars. In response to the HGCRA the JCT is now very specific about the date of issue of Interim Certificates. The first interim certificate should be issued within one month of the date of possession and all subsequent interim certificates should be issued on the same date each month until the date of practical completion. Inevitably one or more of these dates will fall at a weekend or on a Bank Holiday in which case the interim certificate is to be issued on the date of the nearest business day. If the Employer fails to fill in this section of the Contract Particulars then by default interim certificates must be issued at regular periods, not exceeding one month, up until the date of practical completion. After the date of practical completion interim certificates are to be issued as and when monies become due to the contractor, although the architect cannot be required to issue an interim certificate within one month of a previous interim certificate. Specific attention is drawn to the fact that an interim certificate must be issued after the later of the two following events: the end of the rectification period or the issue of the certificate of making good (see clause 4.9.2). The main purpose of this interim certificate is to release the final portion of retention to the contractor.

From the contractor's point of view, the exact position of the day in the month when an interim certificate is to be issued may well affect his cash flow as, for example, his invoices for materials become payable at the end of the month. Also, if the formula method of recovery of fluctuations applies (see p. 169), the fluctuations amount included in a valuation will be affected by the relationship between the 'valuation date' and the date of publication of the monthly indices.

From the Employer's point of view the 'interim certificate date' will affect his cash flow situation also. If the Employer is a public authority and it is necessary for payments to be approved by a finance committee, then the date of the committee meeting in each month may have a direct bearing upon the most appropriate date for the issue of interim certificates, if payment is not to be delayed unduly.

The architect usually relies on the quantity surveyor to advise the sum which should be stated as due in an interim certificate. In order to do this, the surveyor prepares a valuation in accordance with the provisions of clause 4.16 of the JCT Form. As an alternative to clause 4.16 the JCT recognizes that some Employers may wish to make interim payments based upon the principle of 'stage payments'.[2] Interim certificates will be issued at the normal monthly period and payments will be calculated on the basis of stages in the construction work that have been completed. Thus, the first interim certificate may contain the value for the completed substructure and ground slabs, the second certificate the value of the superstructure walls and upper floors, the third certificate the value of the roof, and so on. Stage payments are more appropriate to housing contracts than to more complex projects

Table 10.1 Typical breakdown of house type contract sums for use in 'stage payments' method of interim valuations

Stages	*Proportion of contract value for house types*				*Totals*
	A (£)	*B (£)*	*C (£)*	*D (£)*	
Sub-structure	9,200	9,920	9,568	11,408	40,096
External walls, upper floors, windows and external doors	27,120	31,472	28,656	35,424	122,672
Roof, internal walls and partitions, first fixings	23,496	26,880	25,024	31,816	107,216
Plumbing, internal finishings, second fixings	29,944	33,328	31,616	33,728	128,616
External works	9,920	11,648	11,080	17,440	50,088
Preliminaries	14,800	15,856	15,352	18,816	64,824
Totals	114,480	129,104	121,296	148,632	513,512

where the stages are often not so readily or satisfactorily definable. Table 10.1 shows how a housing contract, consisting of four house types may be broken down into stages, in preparation for interim valuations.

Where the interim valuation is to be prepared on the basis of clause 4.16 it will be necessary for the surveyor personally to visit the site to see for himself both the completed work and the unfixed materials and goods and then to ascertain their total value. The surveyor is required to carry out this task in the seven-day period leading up to the date of issue of the interim certificate (see clause 4.16). This should help to maximize the value of the interim valuation and aid the contractor's cash flow.

For administrative purposes, it will be advantageous to decide from the start a fixed date in each month on which to do valuations. Although the surveyor has no obligation under the terms of the JCT Form to notify the contractor when he proposes to visit the site for valuation purposes, nevertheless it is desirable as well as courteous to do so and even better if the surveyor and the contractor's surveyor agree on a 'valuation date' in each month.

General procedure

It is universally accepted that interim payments are approximate only and, provided the amounts included for the various constituents are reasonable, no objection will be raised because they are not exact. For this reason, the amounts shown in certificates are always round pounds, as also are the sums in the main money column of valuations.

The proper way to prepare an interim valuation is to value, on each occasion, the amount of work which has been done since the beginning of the contract and the value of unfixed materials and goods on the site on the 'valuation date'. From the total value so arrived at, the total of previous payments on account, if any, will be deducted, leaving a balance due for payment. By following this procedure strictly, any undervaluation or overvaluation of either work completed or of unfixed materials on the last previous 'valuation date' will be automatically corrected. The surveyor should not attempt to value in isolation the work done and the materials delivered since the last valuation.

Experience will confirm that valuations are most conveniently set out on double billing paper, thus allowing for sub-totals to each section and subdivision being carried over into the right-hand money column (see example on p. 221). Such sub-totals are often useful for carrying forward into subsequent valuations.

Preparing the valuation on site

The surveyor's first task on visiting the site for valuation purposes (having first made his presence known to the site agent), is to tour the Works, making notes as necessary of the extent of work done and listing the quantities of the various materials and goods stored on the site. He will then be in a position to get down to preparing a draft valuation in the site agent's or clerk of works' office (if there is no office for the surveyor's exclusive use, as often there is not). It is highly desirable, if at all possible, that the valuation total be determined before leaving the site.

As already stated, the surveyor and contractor's surveyor will usually agree to meet on site at a mutually convenient time on the 'valuation date' in the month and will prepare the valuation together, although, of course, the responsibility for the resulting recommendation to the architect will be that of the surveyor alone. Thus, the total sum and its constituent amounts will normally be agreed before leaving the site and subsequent dispute will be avoided. Sometimes, the contractor's surveyor is happy to leave valuations entirely to the surveyor and will accept whatever amounts the architect certifies.

Some contractors prefer to prepare detailed applications for interim payments themselves which they submit to the surveyor a few days before the valuation date. The JCT has acknowledged this practice through clause 4.12, whereby a contractor is entitled to submit an application, to the surveyor, detailing what he considers the gross valuation should be. If a contractor wants a surveyor to consider his application it must be submitted at least 7 days before the date of the interim certificate. On receiving such an application the surveyor is obliged to carry out an interim valuation and if he finds he disagrees with the amounts in the contractor's application he must provide the contractor with a statement detailing the area of disagreement. The surveyor is to provide the contractor with such a statement at the same time as he prepares his interim valuation. Where a statement of disagreement is provided it must be prepared to the same level of detail as was contained in the contractor's original application. This proviso should provide contractor's with an incentive to produce reasonably detailed applications, if they wish, and should reduce payment disputes.

Even when such detailed applications are not made, it may be very helpful to have a list, prepared beforehand by the site agent, of the quantities of unfixed materials stored on site. Surveyors often arrange with contractors for such a list to be available when they arrive on site to prepare the valuation. The surveyor should then check the list, including the quantities, when making the initial tour of the Works.

Inclusions in valuations

Clause 4.10 of the JCT Form provides that 'the amount stated as due in an Interim Certificate shall be an amount equal to the Gross Valuation' of specified constituent parts of the Works, less the Retention Percentage and the total of previous interim certificates. In order to determine 'the gross valuation', the value of each of the constituent parts has to be ascertained. Those parts may be any or all of the following:

1 Preliminaries
2 Main contractor's work (as billed)
3 Variations
4 Unfixed materials and goods
5 Statutory fees and charges
6 Fluctuations in costs of labour, materials and/or taxes, etc.
7 Retention
8 Claims for direct loss and/or expense (see Chapter 9).

Items 1 to 7 on the above list will now be considered in detail. The examples at the end of the chapter illustrate the application of most of the principles and procedures discussed.

1. Preliminaries

Where there is a bill of quantities, the evaluation of the Preliminaries will usually be less difficult than where there is not. In the latter case, it will be necessary, before the date of the first valuation, to make an apportionment of the contract sum to each of its principal constituents. The Preliminaries constituent will then need to be broken down into separate amounts such as would normally be seen in the Preliminaries section of a bill of quantities. This process will be very much easier and the results will be more satisfactory, if there is a Contract Sum Analysis (see p. 105) which shows the total of Preliminaries included in the tender. The contractor will be able to produce the build-up of the tender and the individual amounts allowed in it for the main Preliminaries items. These amounts will then be available for valuation purposes – and later, for the adjustment of Preliminaries, if necessary, in the final account – just as if there had been a bill of quantities.

Even where there is a bill, the situation may still present difficulties. The contractor may have put amounts against all, some, one or none of the priceable items in the Preliminaries section, having included the balance of their value, if any, in other parts of the bill. The surveyor can only take account, of course, of those items which have been priced, even though they may represent only part of the total cost of the Preliminaries.

It is not uncommon to deal with the total value of the Preliminaries as if the items were either all cost-related or all time-related.[3] Thus, in the first case, the total of the Preliminaries is calculated as a percentage of the contract sum after deducting the Preliminaries. In each valuation, this percentage is applied to the total value of item 2 of the list on p. 206. In the second case, the total of the Preliminaries is divided by the contract period (in months) and the resulting sum is multiplied by the number of months which have elapsed to date. This amount is then included in the valuation as the total amount for Preliminaries.

There is no practical reason why the Preliminaries should not be dealt with in either of these two ways, although it should be recognized that they give only approximate results. It is argued in their defence that these methods save time, although it should be pointed out that once the allocation of amounts to the four categories as illustrated in Table 10.2 has been done, it takes very little longer each month to calculate more nearly the true value.

The objection to both these methods is that they result in overpayment of some items and underpayment of others. Also, where the time-related method is used and the contract runs behind schedule, there will be a danger of overpayment, unless the fixed monthly amounts are adjusted. Where the cost-related method is used, there is a danger of inadvertently exceeding the total of the Preliminaries when the total value of the contractor's own work has been significantly increased by variations.

None of the foregoing objections matters very much when the contract is keeping to schedule and no real problems arise. They will matter, however, if the employment of the contractor is terminated because of insolvency. Then, the likelihood will be that the Preliminaries will have been overpaid, particularly if the contract is in its early stages. The possibility of such embarrassment to the surveyor (to put it at its best) will be avoided if the small amount of extra time and trouble is taken to ascertain more exactly the value of the Preliminaries.

It will help in arriving at realistic valuations if the sums which have been inserted in the Preliminaries section are analysed, their constituent parts being dealt with as described in the following paragraphs. Again, such analyses will be more satisfactory if provided by or agreed with the contractor, preferably before the contract is signed.

Table 10.2 Three-storey block of flats, Thames Street, Skinton. Breakdown of Preliminary items*

Items	*Time related (£)*	*Fixed charges (£)*		*Cost related (£)*	*Total (£)*
		(a)	*(b)*		
Management and staff	75,000	–	–	–	75,000
Site accommodation	4,357	1,500	500	–	6,357
Electric lighting and power	900	–	–	–	900
Water	–	117	–	1,260	1,377
Telephones	450	150	50	–	650
Safety, health and welfare	4,380	2,250	1,030	–	7,660
Cleaning	–	–	2,770	–	2,770
Drying out	–	–	2,250	–	2,250
Security	5,450	–	–	–	5,450
Small plant and tools	–	100	250	2,700	3,050
Earthmoving plant	3,300	400	300	–	4,000
Piling plant	4,700	200	100	–	5,000
Temporary roads	1,350	2,000	1,000	–	4,350
Access scaffolding	2,250	1,350	850	–	4,450
Fencing and hoardings	870	3,000	750	–	4,620
Insurance against injury	–	–	–	10,000	10,000
All risks insurance	–	–	–	12,920	12,920
Total	103,007	11,067	9,850	26,880	150,804

Notes: Total of time-related amounts = £5150 per month for the Contract Period of 20 months.

Fixed charges (a) are sums expended at or soon after the commencement of the Contract for delivery, erection, installing, etc.

Fixed charges (b) are sums expended mainly towards the end of the Contract for dismantling and removal.

Total of cost-related amounts = say 2.15% of the Contract Sum less Preliminaries and provisional sums.

*See Appendix A for full details of the Contract.

Preliminaries items (or their components) are of four kinds, namely, cost-related, time-related, single payment or a combination of two or more of the others. Those which are cost-related (e.g., Water for the Works) depend for their value on that of the remainder of the contract sum or of its labour content. Time-related items (e.g., site supervision) depend for their value on the contract period. Single-payment items (e.g., the provision of temporary access roads) are those whose value is not affected either by the value of the rest of the contract or by the contract period but are carried out at a particular point in the progress of the Works. The cost of some Preliminaries items consists of one or two single-payment components and a time-related one, e.g., the provision of a tower crane, involving single payments for erection and dismantling and a weekly hire charge for the intervening period.

The surveyor's next task prior to the first valuation, after having ascertained, where appropriate, the breakdown of the Preliminaries items which have been priced, is to put them into the categories described above, so that the total of each category is then readily available for each valuation thereafter. Table 10.2 illustrates how this may be done. See p. 133 with regard to the rules in SMM7 for the pricing of Preliminaries items which will facilitate their breakdown into the categories shown in Table 10.2.

Even this slightly more detailed approach to the valuation of preliminaries has its drawbacks as it assumes that all the time-related charges are evenly spread throughout the contract period and the fixed charges occur conveniently at the start and end of the contract. However, this will not always be the case. For example, the tower crane mentioned previously may not come onto site until the start of month three and it is then dismantled and removed in month nine. This would result in the contractor incurring fixed charges that occur neither at the start nor at the end of the project, and time-related charges that run for a period of six months and not the full 12-month period. The only true way of trying to accurately reflect a contractor's preliminary costs is to value each item individually on merit. This would be a far easier task for the contractor's surveyor, as he would have access to the original programme, method statement and estimator's costings. Therefore, on a large complex project where there is a high value of preliminary items it may be advisable for the contractor and surveyor to agree a more detailed schedule based upon the contractor's method statement.

In the following valuation examples, the data from Table 10.2 is used to arrive at a value for the preliminaries section.

2. *Main contractor's work*

The value of the work carried out by the contractor's own workforce will be readily ascertainable from the measured work sections of the bills of quantities or, in the case of lump sum contracts without quantities, from the Contract Sum Analysis, if provided, or from a schedule drawn up on the basis of an analysis of the tender. The prices contained in the bills of quantities, being 'the Contract Bills',[4] must be used, of course, in the valuation of the main contractor's work, regardless of whether the contractor underpriced or overpriced the work when compiling the tender.

The surveyor, when valuing the main contractor's work, will begin with the 'Groundwork' section (or the 'Demolition/Alteration/Renovation' section if there is one) and proceed in order through all the succeeding work sections which contain items of work which have been wholly or partly carried out. He should put down separately the total valuation of such work within each work section or sub-section, indicating by use of bill item references what is included in each amount. The reason for so doing is that every amount can then be verified and substantiated subsequently, should any query arise. Also, such detail will often prove useful when doing the next valuation. Of course, if the whole of a work section has been completed, it is only necessary to show its total. The examples of valuations given at the end of the chapter illustrate the procedure.

The notes made during the initial tour and inspection of the Works will again be used in assessing approximately the value of any partly completed items, sections or sub-sections. The completed work as a percentage of the whole item should be indicated opposite the bill reference. Actual measurement of the work on site will seldom, if ever, be necessary solely for interim valuation purposes, except in the case of contracts not based on bills of quantities.

As an alternative to using a bill of quantities for valuation purposes it is possible to make use of a priced activity schedule where one is provided. The second recital in the Articles of Agreement requires the contractor to attach a priced activity schedule to the contract bills, unless this requirement has been deleted by the Employer. The schedule would break the whole project down into readily identifiable activities, similar to the stage payments previously described, and a value will be set against each of these activities. The total value of the schedule should equal the contract sum less the value of any work identified as 'approximate quantities', the value of provisional sums and any prime cost sums and associated profit that may appear in the contract bills. For valuation purposes it will be necessary to identify which activities have been completed and value them in accordance with the activity schedule, identify which activities have been commenced but not yet completed and value them as a proportion of the completed activity.

3. Variations

Very few projects, if any, have no variations. On some jobs they number in double or even treble figures. They present a difficulty in the context of interim valuations for several reasons. First, clause 4.16.1.1 of the JCT Form requires that effect be given in Interim Certificates to the value of variations agreed between the Employer and contractor, or that have been valued by the surveyor under the 'valuation rules', or for which a schedule 2 quotation has been accepted. For this to be done, each variation must be valued as soon as possible after issue. Pressure of other work may cause delay in dealing with variations but it may be dangerous to have to resort to guesswork in consequence. Traditionally contractors and surveyors have tended to accept an approximate evaluation of the net effect of variations when preparing an interim valuation, but a strict interpretation of clause 4.16.1.1 is that no variation should be included unless its value has been agreed between Employer and contractor or valued by the surveyor under the valuation rules. It is wise, therefore, to give priority to the task of valuing variations before each valuation date. The methods of valuing variations are discussed in Chapter 7.

Secondly, the question of how best to deal with the effect of omissions on the measured sections of the bills of quantities is bound to arise. There are two alternatives. The first is to ignore the variations when dealing with the work measured in the bills, the omissions being allowed for when adding the net value of variations. The second is to take the omissions into account when valuing 'main contractor's work' and adding the value of additions only against the 'variations' subhead.

In practice, a combination of the two alternatives will serve best. The first may be used as a general rule, the second being used where a complete section or group of items in the bills is affected by a variation. For example, a variation increasing the width of some (but not all) of the windows will affect part of the items for lintels, cavity trays, sills, window boards, etc. In such a case, it will probably be simpler to include the full value of the items as originally billed in the value of 'main contractor's work', leaving the net value of the variation to take care of the omissions.

If, however, a variation is ordered changing the specification of all the internal doors from painted plywood-faced to hardwood-veneered and the linings and architraves from softwood to hardwood, then the second alternative will probably be preferable, that is, to exclude the whole of the appropriate sub-sections of the Windows/Doors/Stairs and Building Fabric Sundries sections of the bills from the value of 'main contractor's work' and include in the 'variations' part of the valuation the total value of the additions items.

4. Unfixed materials and goods

Unfixed materials and goods are categorized under two separate headings, i.e., materials on site and materials off site.

Materials on site

Clause 4.16.1.2 of the JCT Form requires the inclusion in the amount of an interim certificate of materials and goods delivered to the site for incorporation in the Works. In order to value them, the surveyor will need to be satisfied that the materials and goods are actually on the site and to ascertain approximately how much of each material or good there is. It is not necessary to know the exact quantities because by the time the next valuation is done, most or all of the materials will have been incorporated into finished work which will then be valued as such.

It is helpful if the contractor's surveyor prepares a list of all the materials with their respective quantities, immediately prior to the valuation date and has supporting delivery notes and invoices available so that each of the materials can be valued at invoice cost. The surveyor will have then only to check that the materials and goods are on site and that the quantities and value are as stated. If not, he will amend the list accordingly. The most recent deliveries will probably not have been invoiced as yet and the costs of the equivalent types and qualities on the latest available invoices will be used instead.

It will be necessary to reduce or exclude the invoice value of (a) insufficiently protected materials or those which have deteriorated or been damaged; (b) quantities which are clearly in excess of requirements; and (c) any materials or goods which have been delivered prematurely, that is, when the length of time before they are likely to be required is unreasonably long.

Materials off site

The JCT Form includes a provision in clauses 4.16.1.3 and 4.17 for the value of materials and goods intended for the Works but not yet delivered to the site to be included where certain criteria have been met. Briefly, these are:

(a) the items must appear in a list prepared by the Employer and supplied to the contractor which is subsequently attached to the contract bills (see definitions in clause 1.1);
(b) the goods, etc., must be in accordance with the contract;
(c) the contractor has provided reasonable proof that he has a good title to the goods which will allow ownership to pass to the Employer, upon payment by the Employer to the contractor;
(d) the contractor is to insure the goods for specified perils from the time they become the property of the contractor until delivery to site (after that date the goods would be covered by the clause 6.7 insurance);
(e) that they are set apart at the place where they are being kept and are clearly identifiable and appropriately marked;
(f) bonds have been provided for listed items and 'uniquely identified listed items', unless this requirement has been deleted in the Contract Particulars. The JCT[5] has provided advice on how to interpret these two different categories of goods. For example, a uniquely identified item could be an item of plant such as an air-handling unit, boiler, lift motor, etc., whereas listed items not uniquely identified would tend to cover general construction materials – bricks, blocks, tiles, copper tube, cable drums, etc.

The surveyor must satisfy himself that all the above conditions have been met before he accepts a claim for the value of the off-site goods to be included in a valuation.

The question whether to allow for overheads and profit on the value of unfixed materials and goods sometimes arises. This may at first appear to be a reasonable suggestion, as the value of each part of the Works is presumed to include those factors and the materials and goods are to become part of the Works. It may appear also as a reasonable interpretation of the words 'total value' in clauses 4.16.1.2 and 4.16.1.3. The Q.S. (Practice and Management) Committee of the Royal Institution of Chartered Surveyors has expressed the opinion that 'the word "total" is used only to indicate the collective value of work executed and materials delivered ...' and 'could not be interpreted as including anything in addition to their value *for the purpose of the interim certificate*'.[6]

5. Statutory fees and charges

Any sums paid by the contractor to any local authority or statutory undertaker in respect of fees or charges for work executed or materials or goods supplied in the course of carrying out its statutory obligations should be included in the next interim valuation thereafter. Such charges would commonly be for connections to water, gas and electricity mains. Normally, there will be provisional sums included in the bills of quantities in respect of such work.

6. Fluctuations in costs of labour, materials, etc.

The amounts, if any, under this heading which should be included in interim valuations, will vary according to whether the contract is a firm price or fluctuating price one (see p. 155 for definitions of these terms). The subject of fluctuations has been dealt with in detail in Chapter 8 and it will suffice to add only two further points here. First, where the 'traditional' method of recovery is allowed for in the contract conditions, amounts due should be included in interim valuations as early as possible after the costs have been incurred. Second, the amounts of fluctuations calculated by that method are payable in full (that is, they are not subject to retention) with the percentage addition (if any) referred to in Fluctuations Option A.12 or B.13 of the JCT Form and as stated in the Contract Particulars. It should be noted, however, that amounts calculated by the 'Formula Rules' method are subject to retention (see clause 4.16.2.3 of the JCT Form).

7. Retention

It is a common provision in most standard forms of building contract for a percentage of the valuation total to be deducted. Thus, the sum of money deducted is said to be 'retained' by the Employer and is called 'the Retention' in the JCT Form. The percentage used to calculate the amount of the Retention is called 'the Retention Percentage' (see clause 4.20).

The purpose in retaining part of the total value of work completed to date is (a) to provide an incentive for the contractor to complete the Works promptly, and (b) to 'cushion' the Employer to some extent against the effects of the contractor defaulting, should that happen. The actual percentage retained will be 3% unless a different percentage is inserted in the Contract Particulars.

The exclusions from the application of the Retention Percentage where the JCT Form applies are set out in clause 4.16.2. For convenience of reference, they are listed as follows:

(a) additional insurance premiums payable by the contractor as a result of the Employer making early use or occupation of the site or works (clause 2.6.2);
(b) statutory fees and charges (clause 2.21);
(c) costs relating to patent rights (clause 2.23.2);
(d) the cost of opening up work which has been covered up and/or the cost of testing materials, goods or executed work (clause 3.17);
(e) insurance premiums payable by the contractor in order to maintain insurances against damage to property other than the Works (clause 6.5.3);
(f) increase in terrorism insurance premium (Insurance Option A.5.1)
(g) insurance premiums which should have been paid by the Employer but in regard to which he has defaulted (Insurance Options B.2.1.2 and C.3.1.3);
(h) loss or expense due to the regular progress of the Works having been affected by specified matters (clauses 3.24 and 4.23);
(i) insurance payments for restoration work, etc. (Insurance Option B.3.5 and C.4.5.2);
(j) fluctuations in the cost of labour and materials, etc., calculated other than by the 'Formula Rules' method (Fluctuations Option A and B).

The JCT Form provides, in clause 4.20.3, for one-half of the total then retained to be paid to the contractor in the next interim certificate after the architect has issued a certificate of Practical Completion. The amount of money the contractor receives is subject, of course, to any right of deduction the Employer has exercised against the retention (see clauses 4.13.2 and 4.18.4). This release of retention obviously applies where the whole Works has reached practical completion, but it will also apply to 'Sections' of the work that have reached practical completion and where the Employer has used the option to split the works into Sections (see the sixth recital).

The same provision will operate under clause 2.33 of the JCT Form, should the Employer take possession of any part of the Works before the Date for Completion stated in Contract Particulars. For the purpose of retention it is deemed that the portion of the works taken over by the Employer has reached practical completion. As a result the contractor is entitled to receive half of the retention that had been deducted from that portion of the works. For example, if the portion of the works is valued at £200,000 and retention is 3%, the contractor should have £3000 released to him in the next interim certificate.

The second half of the retention monies is released when the contractor has received a Certificate of Making Good. The architect will issue this certificate once he is satisfied that the contractor has made good, at the architect's request, such defects as have appeared during the Rectification Period. A similar provision applies to any Section of work identified in the contract or part of the Works where the Employer has taken early possession.

Retention bond

During the 1990s, a number of construction bodies lobbied the Government and contract drafting organizations, seeking the abolition of retention from construction contracts. The JCT has partially met these demands by including an optional condition (see clause 4.19) whereby a contractor may provide a retention bond in lieu of the Employer deducting retention from the contractor's interim payments. This optional clause is not to be used where the Employer is a local authority (see Chapter 17 for further details).

Notifying the architect

When the surveyor has completed his valuation, he must inform the architect as soon as possible of the amount of the payment recommended. Prompt notification is essential if the architect's certificate is to be issued by the date set out in the Contract

Particulars. The Royal Institution of Chartered Surveyors publishes a standard form for this purpose in sets of three forms (one copy each for architect, contractor and surveyor) in pads of 100 sets. Alternatively computer software enables administration forms to be easily created which can be transmitted electronically and/or as hard copy. An example of such a form is demonstrated in Figure 10.2, completed to correspond with the first of the examples of draft valuations.

The form provides for the totals only of (a) the gross valuation; (b) retention; (c) repayments of any advance payments; (d) amounts included in previous *certificates*; and (e) balance due for payment. Supporting information should be provided specifying the retention deductions made (clause 4.18.2). The detailed build-up of valuations is not normally sent to the architect.

Payments and notices

The date of issue of the interim certificate sets the date when the Employer becomes responsible for making a payment to the contractor, i.e., the 'due date'. The Employer then has 14 days from the due date to make the payment to the contractor and the last day of this 14-day period is the 'final date for payment'. The 'due date' and 'final date for payment' are two of the key requirements of the HGCRA to ensure that a clear payment timetable is set out in the contract.

No later than five days after the issue of the interim certificate the Employer is required to give the contractor a written notice stating:

- What amount he intends to pay of the interim certificate.
- What items the payment relates to.
- The basis on which the amount was calculated.

This notice may be viewed as an unnecessary duplication of the architect's interim certificate but such a notice is a requirement of the Construction Act.[7] In this notice the Employer could inform the contractor what he intends to pay if he disagrees with the amount in the interim certificate but it is thought that in most instances the Employer will pay the amount certified. However, the JCT have largely removed this administrative burden from the Employer through clause 4.13.5, i.e., if the Employer fails to issue a notice the contractor is entitled to the amount stated in the interim certificate. Therefore, an Employer only needs to issue the notice if he intends to pay a different amount from that stated in the interim certificate.

Not later than five days before the final date for payment of an interim certificate the Employer may give a written notice to the contractor specifying any deductions or monies that are to be withheld from the amount due. This may cover deductions that the Employer is entitled to make from monies due, i.e., liquidated damages, employment of other contractors (clause 3.11), insurance premiums (Insurance Option A.2), etc. It would also allow the Employer to hold back monies for defective works that had been discovered since the issue of the interim certificate or for work that had not been properly completed. This notice is commonly referred to as a 'withholding notice'.

Employer's failure to pay

The final date for the Employer to make a payment to the contractor is 14 days from the date of issue of the interim certificate. If the Employer fails to pay the amount by the final date then the contractor is entitled to receive simple interest at 5% above the Bank of England base rate current at that time. The interest period would run from the final date for payment of the interim certificate until the day the payment is made. The interest payments are not to be added on to the interim certificates they are treated separately as a debt due from the Employer to the contractor.

A further right the contractor possesses is that he may suspend his obligation to carry out the works. Where the Employer has failed to make a full payment then the contractor may give a written notice to the Employer (copy to the architect) which provides the following:

- A statement of intention to suspend carrying out the works.
- The grounds on which it is intended to suspend performance.

If the Employer fails to make the payment within seven days of being given the notice, then the contractor may suspend his operations until full payment is received.

A final sanction that the contractor possesses is to terminate his employment. This is allowed under clause 8.9.1.1. However, termination should only be considered where a substantial sum is involved and other reasonably avenues have been tried without success.

Advance payments

There is provision in the contract for a contractor to receive payment in advance (see clause 4.8). The principle is that a lump sum would be paid to the contractor on the date stated in the Contract Particulars. The contractor would then repay the monies in accordance with the schedule in the Contract Particulars. This optional clause is not to be used where the Employer is a local authority but other Employers may wish to make an advance payment where a contractor incurs high costs at the early stages of a project. For example, there may be a long lead-in time from when the contract was signed and the contractor is to be given possession of the site, during which time the contractor is required to carry out design and prefabrication work and purchase items of plant. As the interim payment procedures will not normally commence until after the date of possession an Employer would be able to pay the contractor for this early work through the advance payment option. To provide the Employer with some security it is possible the contractor may be required to provide an 'advance payment bond' (see Chapter 17).

Liquidated and ascertained damages

It should be understood that no deduction should be made from the amount of a valuation for liquidated and ascertained damages (see p. 242) because of failure on the contractor's part to complete the Works on time. Even though the surveyor knows that the Employer is entitled to such damages, there is no provision in the JCT Conditions of Contract for anyone other than the Employer himself to make the deduction from any sum due to the contractor.

The Employer may, if he so wishes, waive his right to damages, in which case he will make no deduction. It may, however, be prudent for the surveyor to remind the architect that damages are due to the Employer – and even how much – so that the architect may be in a position to advise the Employer accordingly. It would also be prudent to remind the Employer of the need to issue a withholding notice, as previously described, before any monies are held back from an interim certificate.

Examples of interim valuations

The following examples illustrate many of the details which have been discussed earlier in the chapter. They relate to a contract of which particulars are given in Appendix A on pp. 384–387.

Valuation No. 1

The first interim certificate is to be issued on 19 October 2005, one month after the contractor took possession of the site. The contractor has not submitted an application for payment on this occasion. The quantity surveyor must not prepare his interim valuation more than 7 days before the date of the interim certificate. Consequently the surveyor arranged to meet the contractor's surveyor on site on 12 October to agree the first valuation (see Figures 10.1 and 10.2). From an initial inspection of the Works, the surveyor found that the following work had been done.

About two-thirds of the demolition work and about 10% of the excavation for the main building; some concrete bases were cast (items 27A–27C in Bill No. 3); about 10% of the drainage work; and the water and electricity companies had submitted charges for preliminary work to the mains connections.

Mess huts, drying sheds, site offices and about half of the temporary fencing erected and temporary water supplies provided; about half of the temporary roads were laid.

Quantities of sand, aggregate and cement stored on site were estimated and noted down.

Commentary on Valuation No. 1

1 As part of the Preliminaries is cost-related, they cannot be calculated until the cost of the builder's work is known. This is indicated by inserting the words 'see below' opposite 'Bill No. 1 – Preliminaries' (Figure 10.1). See p. 208 for breakdown of Preliminaries.

Shops and Flats, Thames St., Skinton

Valuation No. 1 – 12 October 2005			£	£
Bill No. 1 – Preliminaries				*See below*
Bill No. 2 – Demolition	say 66% of 5,531			3,650.00
Bill No. 3 – Shops and flats				
Groundwork	say 10% of £181,837		18,183.00	
In situ concrete 27A–27C			9,801.00	
				27,984.00
Bill No. 4 – External works				
Drainage	say 10% of £16,004			1,600.00
				33,234.00
Addition to correct for errors in bills of quantities			2.395%	796.00
				34,030.00
Preliminaries*				
Time-related items			5,150.00	
Fixed charges			5,530.00	
Cost-related items 2.15% of £34,030			732.00	11,412.00
				45,442.00
Statutory bodies				
Thames Water – water main connection			1,000.00	
Seeboard – electric main connection			1,000.00	2,000.00
Materials on site				
Sand	10 tonnes	£16.50	165.00	
Aggregate	20 tonnes	£22.00	440.00	
Cement	9 tonnes	£108.00	972.00	1,577.00
				49,019.00
Retention			49,019.00	
Less amount subject to nil retention:				
Statutory Bodies			2,000.00	
		3% of	47,019.00	1,411.00
	Total of Valuation No. 1			47,608.00

* see p. 208

Figure 10.1 Valuation No. 1

KEWESS & PARTNERS 52 High Street, Urbiston, Middlesex, UN2 1QS

Employer:	Contractor:	Architect:
Skinton Developments plc High Path Skinton	Beecon plc River Road Skinton	Draw & Partners 25 Bridge Street Skinton

Works: Shops and Flats, 28-34 Thames Street, Skinton

Interim Valuation

Valuation No: 1 | Date of issue of Interim Valuation: 12 October 2005 | Date of issue of Interim Certificate: 19 October 2005

Gross valuation of the Works as at 12 October 2005		49,019.00
Less retention as Statement below		1,411.00
Net valuation		47,608.00
Less reimbursement of advance payment		0.00
Less amounts previously certified up to and including Interim Certificate No. Not applicable		0.00
Balance due	£	47,608.00

Retention Statement

Value of Contractor's work subject to:

Full retention of 3%	47,019.00	1,411.00
Half retention of 1.5%	0.00	0.00
No retention	2,000.00	0.00
Gross valuation	49,019.00	
Total retention calculated		1,411.00

Figure 10.2 Valuation form No. 1

2 For calculation of the percentage adjustment for errors in the bills of quantities, see p. 98. This item is inserted before the 'Preliminaries' are added because they were excluded when calculating the percentage.

Valuation No. 5

Interim certificate number 5 is due for issue on 19 February 2006 which is a Sunday, this means the certificate must be issued on the nearest business day, i.e., Monday, 20 February. On this occasion the contractor submitted an application for payment on 10 February. The quantity surveyor visited the sited on 13 February to gauge the amount of work done to date, and compare that with the contractor's application.

The domestic piling sub-contractor had finished his work and had left the site; excavation was about 80% complete; varying amounts of concrete work and brickwork had been done; woodwork in the roof of the single-storey part was being fixed; some windows, external doors and internal door frames were in position; plumbing and electrical work was in progress; about 45% of the drainage and about 25% of the permanent fencing had been done. Quantities of sand, aggregate, cement, bricks, partition blocks, timber, felt damp proof course and wall ties were stored on site – these were estimated and noted down. The Contract Particulars identified the lift car and associated equipment as a listed item that the Employer was prepared to pay for before delivery to site. The contractor has claimed £20,000 for this item. The quantity surveyor has obtained information confirming that the relevant criteria in clause 4.17 have been complied with.

The quantity surveyor was able to confirm that the contractor's application was an accurate reflection of the work carried out, with the exception that the contractor had incorrectly claimed 90% of the permanent fencing as being complete. Also the contractor had claimed £5000 for sanitary ware stored on site, although these materials were on site the quantity surveyor considered they had been brought onto site far too early and was not prepared to include them in his valuation. The quantity surveyor produced his interim valuation and clearly identified where he had made changes to the contractor's application with reference to the permanent fencing and the sanitary ware.

During the visit, the quantity surveyor and contractor's surveyor agreed the revised total of £1213 for increased costs of labour and materials, which included 10% addition under Fluctuations Option B.13 (see Contract Particulars in Appendix B). Three variation orders had been issued so far but it was agreed that not enough had been done on them to be taken into account in this valuation. The total of Certificates 1–4 inclusive was £269,826.

Commentary on Valuation No. 5

1 As part of the Preliminaries is cost-related, they cannot be calculated until the cost of the builder's work is known. This is indicated by inserting the words 'see below' opposite 'Bill No. 1 – Preliminaries' (Figure 10.3). See p. 208 for breakdown of Preliminaries.
2 For calculation of the percentage adjustment for errors in the bills of quantities, see p. 98. This item is inserted before the 'Preliminaries' are added because they were excluded when calculating the percentage.
3 The fixed charges part of Preliminaries is the total of 'Fixed charges (a)' on p. 208.
4 The deduction for previous payments should be the total of the architect's certificates, not of the surveyor's valuations, as it is possible for them to differ.
5 The contractor had received an advance payment of £50,000 from the Employer (see Contract Particulars) mainly for some early mobilization costs and costs associated with the lift design and prefabrication works. This sum is to be repaid to the Employer on 20 February 2006. The amount of the advance payment is to be deducted from the gross valuation (see clause 4.10.2) and should be deducted after the retention has been calculated.

Shops and Flats, Thames St., Skinton

Valuation No. 5 – 13 February 2006			£	£
Bill No. 1 – Preliminaries				*See below*
Bill No. 2 – Demolition				5,531.00
Bill No. 3 – Shops and Flats				
Groundwork	say 80% of	£181,837	145,470.00	
In situ concrete	p. 27	24,755		
	28A-28C	12,985		
	29F-29H	11,100		
	30A-30D	6,888		
			55,728.00	
Masonry	35A-36F	31,618		
	37C-37E	5,380		
	42A-42D	7,460		
			44,458.00	
Structural/timber	47E-47H	2,865		
	52A-53F	3,364		
			6,229.00	
Waterproofing	80A-81C	1,400		
	89C-90A	2,418		
			3,818.00	
Winds/doors/stairs	93A-94D	6,664		
	98B-98C	7,283		
	99D-99G	4,878		
			18,825.00	
Surface finishes	105A-106C	12,318		
	108D-108E	937		
			13,255.00	
Piped supply system	125A-126C	6,167		
	127C-127H	662		
	128A-128F	1,635		
			8,464.00	
Electrical services	say 30% of	£32,051	9,615.00	
				311,393.00
Bill No. 4 – External works				
Drainage	say 45% of £16,004		7,202.00	
Fencing	say 25% of £17,244		4,311.00	
				11,513.00
				322,906.00
Addition to correct for errors in b/q			2.395%	7,734.00
Sub-total				342,153.00
Preliminaries				
Time-related items	5 × £5,150		25,750.00	
Fixed charges			11,067.00	
Cost-related items	2.15% of £342,153		7,356.00	
				44,173.00
	Carried forward			386,326.00

Figure 10.3 Valuation No. 5

Valuation No. 5 (Cont.)			£	£
		Brought forward		386,326.00
Statutory bodies				
Thames Water – water main connection			1,000.00	
Seeboard – electric main connection			1,000.00	
Segas – gas main connection			2,000.00	
				4,000.00
Materials off-site lift equipment			20,000.00	
Materials on site				
Sand	20 tonnes	£16.50	330.00	
Aggregate	30 tonnes	£22.00	660.00	
Cement	4 tonnes	£108.00	432.00	
Flettons	16 m	£180.00	2,880.00	
Partition blocks	90 m^2	£9.00	810.00	
Damp proof course	20 m	£1.20	24.00	
Wall ties – stainless steel	400 no.	£0.20	80.00	
Softwood	200 m	£2.35	470.00	
				25,686.00
Fluctuations as agreed*				1,213.00
				417,225.00
Retention			417,225.00	
Less amounts subject to nil retention:				
Statutory Bodies		4,000		
Fluctuations		1,213		
			5,213.00	
		3% of	412,012.00	12,360.00
Total of Valuation No. 5				404,865.00
Less total of Certificates Nos. 1–4				269,826.00
Total amount due				135,039.00

* Includes 10% addition as Fluctuation Option B.13

Figure 10.3 continued

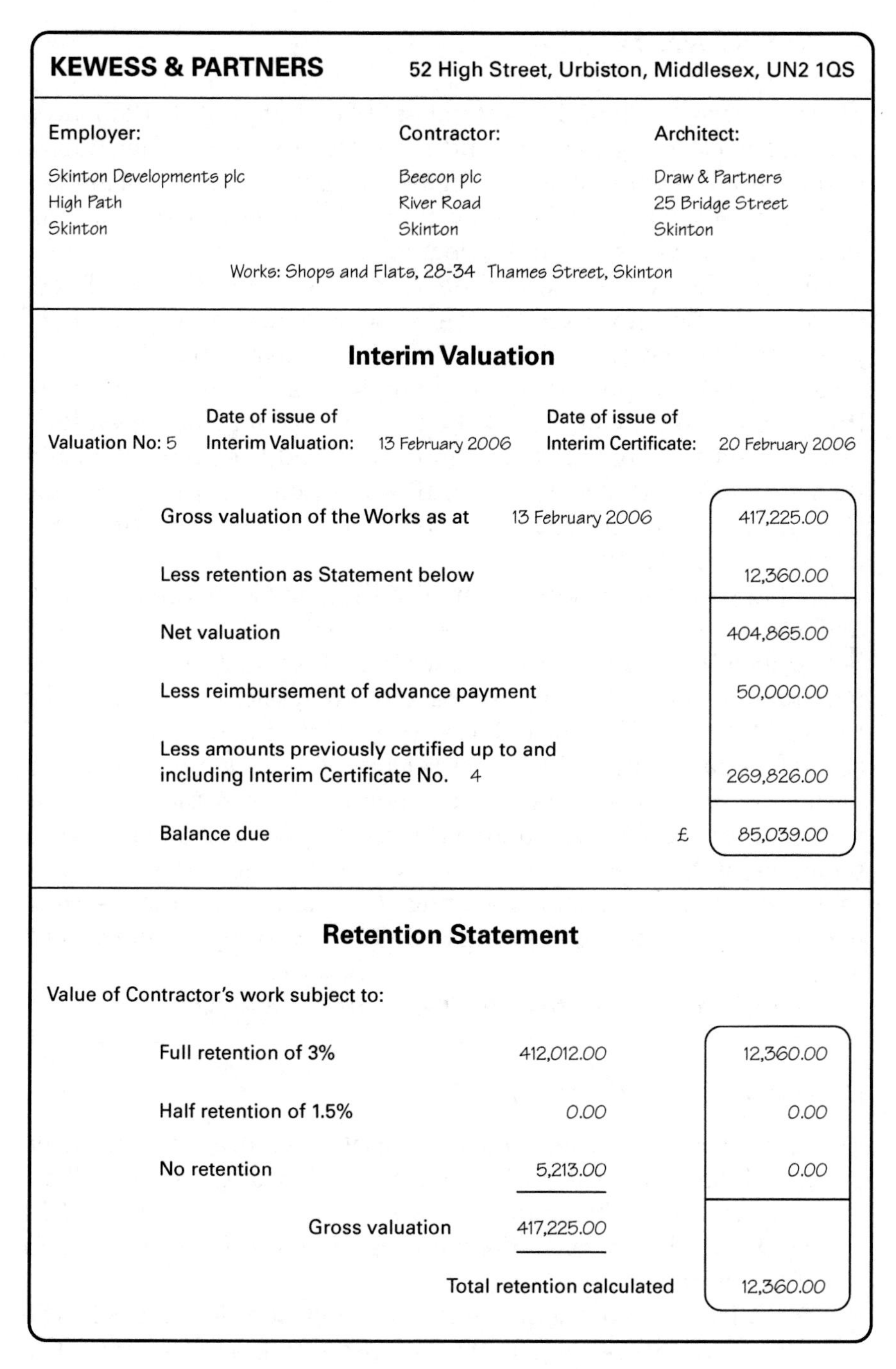

KEWESS & PARTNERS 52 High Street, Urbiston, Middlesex, UN2 1QS

Employer:	Contractor:	Architect:
Skinton Developments plc High Path Skinton	Beecon plc River Road Skinton	Draw & Partners 25 Bridge Street Skinton

Works: Shops and Flats, 28-34 Thames Street, Skinton

Interim Valuation

Valuation No: 5 Date of issue of Interim Valuation: 13 February 2006 Date of issue of Interim Certificate: 20 February 2006

Gross valuation of the Works as at 13 February 2006		417,225.00
Less retention as Statement below		12,360.00
Net valuation		404,865.00
Less reimbursement of advance payment		50,000.00
Less amounts previously certified up to and including Interim Certificate No. 4		269,826.00
Balance due	£	85,039.00

Retention Statement

Value of Contractor's work subject to:

Full retention of 3%	412,012.00	12,360.00
Half retention of 1.5%	0.00	0.00
No retention	5,213.00	0.00
Gross valuation	417,225.00	
Total retention calculated		12,360.00

Figure 10.4 Valuation form no. 5

Valuation No. 12

This was carried out on 14 September 2006, that is, after 12 months had elapsed (see Figures 10.5 and 10.6). Due to bad weather, delays in deliveries because of strikes in the haulage industry during the early part of the year and a delay in installing the lifts, the contract was about five weeks behind schedule.

The five flats on the second floor had been recently completed, those on the first floor and the staircase were in the early stages of plastering, whilst the shops on the ground floor were at first fixings stage. Drainage was finished and the fencing was about two-thirds complete. About 10% of the external pavings had been laid.

Increased costs of labour and materials had been agreed within the past week at £2364, including 10% addition under Fluctuations Option B.13. One variation had been issued since the previous valuation, i.e., AI No. 42.

On 4 September, by agreement with the contractor, the Employer had taken possession of the five flats on the second floor. On 6 September, the architect issued a written statement identifying the part taken over and giving the date of possession, as required by clause 2.33 of the JCT Form. The surveyor estimated the total value of the five flats to be £250,000 for release of retention purposes.

During July a claim had been made by the contractor for loss arising under clause 4.23, due to the architect's failure to provide a schedule of the decorations and details of selected wallpapers required in the flats, thus delaying the completion of the second floor flats. The claim, amounting to £1610, was accepted and agreed on 22 August.

The total of Certificates Nos. 1–11 was £738,432.00.

Commentary on Valuation No. 12

1 As part of the Preliminaries is cost-related, they cannot be calculated until the cost of the builder's work is known. This is indicated by inserting the words 'see below' opposite 'Bill No. 1 – Preliminaries' (Figure 10.5). See p. 208 for breakdown of Preliminaries.
2 For calculation of the percentage adjustment for errors in the bills of quantities, see p. 98. This item is inserted before the 'Preliminaries' are added because they were excluded when calculating the percentage.

Shops and Flats, Thames St., Skinton

			£	£
Valuation No. 12 – 14 Sept. 2006				
Bill No. 1 – Preliminaries				*See below*
Bill No. 2 – Demolition				5,531.00
Bill No. 3 – Shops and Flats				
Groundwork	say 90% of	£181,837	163,653.00	
In situ concrete	pp. 27–30	88,322		
	31A-31E	31,548		
	32C	5,077		
			124,947.00	
Masonry	pp. 34-40	90,926		
	42A-42F	16,034		
	43A	1,003		
			107,963.00	
Structural/timber	47A-47H	14,623		
	50A-52J	12,817		
	53C-58F	14,699		
			42,139.00	
Cladding/covering	70A-73C	13,515		
	78A-79D	4,754		
			18,269.00	
Waterproofing	80A-86C	11,769		
	89C-91B	9,945		
			21,714.00	
Winds/doors/stairs	93A-96C	45,928		
	98B-98D	5,138		
	99D-100B	12,244		
			63,310.00	
Surface finishes	say 55% of	£52,447	28,846.00	
Furniture/equipment	say		2,914.00	
Bldg. fabric sunds.	113A-114D	4,747		
	115A-116A	15,604	20,351.00	
Disposal systems	say 50% of	£29,137	14,568.00	
Piped supply system	say 40% of	£21,045	8,418.00	
Mechanical heating	say 20% of	£84,215	16,843.00	
Security	say 20% of	£22,326	4,465.00	
Electrical services	say 10% of	£32,052	3,205.00	
Lift installation	100%		71,235.00	
				712,840.00
Bill No. 4 – External works				
Drainage	138A-140F		16,004.00	
Fencing	say 60% of	£17,244	10,346.00	
Pavings	say 10% of	142A-142K	804.00	
				27,154.00
				745,525.00
Addition to correct for errors in b/q			2.395%	17,855.00
			Carried forward	790,534.00

Figure 10.5 Valuation No. 12

Valuation No. 12 (cont.)

			£	£
	Brought forward			790,534.00
Preliminaries*				
Time-related items†		12 × £4,905	58,860.00	
Fixed sums (a)		11,067		
(b)		2460	13,527.00	
Cost-related items	2.15% of	£790,534	16,996.00	89,383.00
Variations	as Valuation No. 11		11,468.00	
	AI No. 42		789.00	12,257.00
Statutory bodies:				
Thames Water	Water main connection		5,000.00	
Seeboard	Electric main connection		4,000.00	
Segas	Gas main connection		3,000.00	
Urbiston Council	Sewer connections		1,500.00	
BritishTelecom	Telephone		500.00	14,000.00
Materials on site:				
Cement	8 tonnes	£108	864.00	
Fletton bricks	2 m	£180.00	360.00	
Partition blocks	60 m^2	£9.00	540.00	
Doors	50 No.	£17.00	850.00	
Plumbing goods say			1,200.00	3,814.00
Fluctuations, as agreed‡				2,364.00
Claim under clause 4.25, as agreed				1,610.00
				913,962.00
Retention			913,962.00	
Less amounts subject to nil retention:				
Statutory bodies		14,000		
Fluctuations		2,364		
Claim		1,610	17,974.00	
			895,988.00	
Less amounts subject to half retention:				
	Flats handed over 12/09/06:		250,000.00	
			645,988.00	
	3% of	645,988	19,380.00	
	1.5% of	250,000	3,750.00	23,130.00
Total of valuation No. 12				890,832.00
Less total of Certificates Nos. 1–11				738,432.00
Total amount due				152,400.00

* See page ***
† Reduced from £5150 to allow for one month's delay
‡ Including 10% addition as Fluctuation Option B.13

Figure 10.5 continued

KEWESS & PARTNERS 52 High Street, Urbiston, Middlesex, UN2 1QS

Employer:	Contractor:	Architect:
Skinton Developments plc High Path Skinton	Beecon plc River Road Skinton	Draw & Partners 25 Bridge Street Skinton

Works: Shops and Flats, 28-34 Thames Street, Skinton

Interim Valuation

Valuation No.: 12 Date of issue of Interim Valuation: 15 September 2006 Date of issue of Interim Certificate: 19 September 2006

Gross valuation of the Works as at 14 September 2006		913,962.00
Less retention as Statement below		23,130.00
Net valuation		890,832.00
Less reimbursement of advance payment		50,000.00
Less amounts previously certified up to and including Interim Certificate No. 11		738,432.00
Balance due	£	102,400.00

Retention Statement

Value of Contractor's work subject to:

Full retention of 3%	645,988.00	19,380.00
Half retention of 1.5%	250,000.00	3,750.00
No retention	17,974.00	0.00
Gross valuation	913,962.00	
Total retention calculated		23,130.00

Figure 10.6 Valuation form No. 12

3 The monthly amount of the time-related portion of the Preliminaries has been recalculated to allow for one month's delay.
4 The fixed charges part of the Preliminaries is the total of 'fixed charges (a)' and approximately 50% of the amounts for 'drying out' and 'cleaning' in col. (b) in respect of the part handed over on 4 September 2006, and remaining sections that are nearing completion (see p. 208).
5 The amounts for statutory bodies are based on the fees incurred by the contractor to date and the first four accounts for Employer's telephone call charges that had been received from British Telecom.
6 The deduction for previous payments should be the total of the architect's certificates, not of the surveyor's valuations, as it is possible for them to differ.

References

1. The Cameron McKenna Construction Law Group, Special Report, April 1998.
2. JCT, *Standard Building Contract, 2005 Edition*, clause 4.10.
3. TURNER, Dennis F., *Quantity Surveying Practice and Administration* (London: George Godwin Ltd., 1983), p. 159.
4. JCT, *Standard Building Contract, 2005 Edition*, Articles of Agreement, Second Recital.
5. JCT, Amendment 18 and Guidance notes (London: RIBA Publications, April 1998), note 14, p. 41.
6. The Query Sheet, *Chartered Surveyor, BQS Quarterly*, Vol. 1, No. 2, December 1973, p. 40.
7. JCT, Amendment 18 and Guidance notes (RIBA Publications, April 1998), note 12, p. 40.

11

Final accounts

To be able to make final payment to a contractor of the sum to which he is entitled under the terms of the contract, it is usually necessary to produce a final account. In any but the very smallest jobs, there will be many adjustments to be made to the Contract Sum and a detailed document is necessary to show, to the satisfaction of all concerned, what amount of money the Employer is liable to pay and the contractor is entitled to receive and how it has been calculated.

In the case of lump sum contracts, therefore, the final account begins with the Contract Sum, shows what amounts are deducted and what amounts are added (and for what reasons), and ends with the adjusted total sum. This sum, when agreed by the contractor, is the total amount which the Employer will pay to the contractor for the work he has done.

In the case of measured contracts, the final account is built up from nil to an 'ascertained final sum', which is the aggregate of amounts for named parts of the project. The ascertained final sum is the total amount, which the Employer will pay to the contractor.

The responsibility for preparing the final account is that of the quantity surveyor and the Joint Contracts Tribunal (JCT) Form limits the time in which the task should be completed. As previously stated (see p. 117), the surveyor has a duty to endeavour to complete the variation accounts within the period stated in clause 4.5, and the whole account in time for the architect to issue the Final Certificate at the time specified in clause 4.15.1.

It will be necessary for the surveyor to obtain the contractor's agreement to the total of the account. To be able to give this, the contractor will need to see the supporting details. Accordingly, a photocopy of the whole document, fully priced and totalled, should be sent to the contractor, either complete or in sections as they are finished. As there will often be some points of disagreement, the surveyor and the contractor's surveyor will need to meet and discuss those points, negotiate revisions to any suggested item rates and prices, etc., until eventually the whole is agreed. Each of them (or their principals) will then sign the summary page to signify their agreement.

Constituents of final accounts

Lump sum contracts (except those based on bills of approximate quantities)

The final account will usually consist of a document which contains most or all of the following sections, although not necessarily in the given order.

(a) A summary of the account.
(b) Adjustments of prime cost sums.
(c) Adjustments of Provisional sums.
(d) Variation accounts.
(e) Adjustment of approximate quantities.
(f) Claims.
(g) Fluctuations in costs of labour, materials and statutory contributions, levies and taxes.

The document may be handwritten or it may be printed, depending on the client's requirements. The foregoing lists of sections of the account require comment as follows.

Summary of the account

The summary usually consists of a single page of double billing paper showing the contract sum, the respective total deduction and addition amounts transferred from each of the succeeding sections, and the final total. At the bottom of the page will appear the signatures of the contractor and surveyor. A suitable format is shown in Figure 11.1.

Shops and Flats, Thames St., Skinton
for
Skinton Development Co.
FINAL ACCOUNT

	Omissions £	Additions £
Contract sum as contract dated 19 August 2005		1,448,000.00
Adjustment of provisional sums	50,668.00	52,794.54
Variations	119,587.43	138,256.87
Remeasurement of approximate quantities	849.09	745.32
Fluctuations in costs of labour and materials		5157.61
Claim for loss under clause 4.25		11,610.00
	171,124.52	1,656,564.34
Less omissions		171,124.52
Total of final account		1,485,439.82

Signed for and on behalf of
Beecon Ltd,
River Road,
Skinton,
Middlesex.

Signed for and on behalf of
Kewess & Partners
52, High Street,
Urbiston,
Surrey.

....

11 December 2007

Figure 11.1 Final account summary

Adjustments of prime cost sums

Under the SMM7 measurement rules, a Prime Cost sum may be used when measuring decorative papers or fabrics (Section M52). The measurement rules also stipulate that where a nominated sub-contractor or nominated supplier is to be employed, this should be allowed for in the bill of quantities through the use of a Prime Cost Sum (Rules A51 and A52). However, the JCT 05 edition has had all references to nominated sub-contractors and suppliers removed from its conditions. Therefore, a situation arises where the measurement rules make specific allowance for nomination whereas the contract conditions make no provision for nomination. If an employer did wish to appoint a nominated sub-contractor, this would be allowable under the measurement rules although the contract conditions would have to be amended appropriately.

Despite there being no allowance for nomination, the JCT does require that prime cost sums and contractor's profit be adjusted during the preparation of the final account (see clause 4.3.2.1). Any prime cost sums will have to be deducted along with any contractor's profit to be replaced with the actual costs incurred plus the contractor's profit.

Adjustments of Provisional sums

Similar to the above, all Provisional sums are to be deducted from the account. Where the architect has issued instructions concerning the expenditure of a Provisional sum, this will normally be measured and priced in exactly the same way as those in variation accounts, and added to the account.

A Provisional sum is a sum provided for work or services to be executed by a statutory authority, statutory undertaking or as an allowance for dayworks. Under other circumstances a Provisional sum may be used for 'provisional work', i.e., for items that cannot be properly described or measured or where it is uncertain as to whether the item may be required or not. Under SMM7 measurement rules, such provisional work must be classified as either 'defined' or 'undefined' work. Defined work is work, which is not completely designed at the time the tender documents are issued but for which certain specified information can be given.[1] Undefined work is work for which such specific information cannot be given.

It is important for a surveyor and contractor to appreciate the difference between a defined and undefined Provisional sum. Where provisional work is described as 'defined', the contractor is deemed to have made a proper allowance for that work within his construction programme and to have made an appropriate allowance within his preliminaries. The result being that if the contractor is instructed to carry out the work provided for in a defined Provisional sum, he has no right to request an extension of time if he is delayed as a result of carrying out the work. Similarly the contractor cannot claim any extra preliminaries incurred as a result of carrying out the work. The one proviso is that the description provided in the defined Provisional sum is adequate to describe the work to be executed. If a contractor can show that a defined Provisional sum is not an accurate reflection of the work to be executed, then the description must be corrected so that the necessary information is provided (see clause 2.14.1 and 2.14.3). This correction is to be treated 'as a variation' which would then allow the contractor to submit claims for an extension of time and additional preliminaries where appropriate. Where a Provisional sum is classified as undefined, the contractor is not required to make any additional allowance for this work in his tender. Therefore, if the contractor is instructed to carry out work against an undefined Provisional sum he is potentially able to request an extension of time if applicable and is entitled to recover the cost of any additional preliminaries directly associated with the work.

During the course of the project the architect will issue written instructions to the contractor detailing what work he is required to do, which will be paid for out of the Provisional sums (see clause 3.16). When the contract sum is adjusted at the end of the contract, the Provisional sums, like Prime Cost sums, will be deducted and the total value of the work actually carried out will be substituted. In order to ascertain what that total cost is, the work will (if carried out by the contractor) be measured and valued as executed and will be priced in the same way as varied work is priced (see Chapter 7). Such total cost may also include work valued as daywork. Alternatively, the contractor may be asked to submit a Schedule 2 Quotation for carrying out the work and, if his quotation is accepted, the amount tendered will be set against the appropriate Provisional sum in the settlement of the accounts. Where a Provisional sum has been included for works to be executed by a statutory authority the contractor is to be reimbursed the actual fee incurred, i.e., no allowance for profit (see clause 2.21).

Variation accounts

The variation accounts (see p. 110) may commence with a summary showing the total of deductions and additions for each separate variation or, alternatively, the net total saving or extra for each. Following the summary, the supporting details are given consisting of measured items grouped as omissions and additions under each AI. A third format shows only the total net saving or extra for all the variations together, and the presentation of the details is in the form of a bill of omissions and a bill of additions without items being separated under each variation.

Adjustment of approximate quantities

This section of final accounts requires more detailed consideration as, unlike the other sections, it does not form a separate chapter.

When the quantity of an item or group of items of work cannot be accurately ascertained at the time of preparing the tender documents, the item or group of items is distinguished by being described as an 'approximate quantity'. In the case of single items, this may appear in brackets at the end of the description. In the case of a group of items, for example, those for the substructure of a building, the words 'Approximate quantities' will normally precede the group.

In all cases, whether single items or groups of items, it is incumbent upon the surveyor to deduct the value of the approximate quantities from the contract sum and to substitute the value of measured items as actually carried out. This procedure is required by clauses 5.6.1, 5.6.1.4 and 5.6.1.5 of the JCT Form. Even where the quantity of an item as executed is the same as that in the bills, the deduction and addition should still be shown. Otherwise, it may be supposed that the approximate quantity item has been overlooked. The value of the remeasured items will be ascertained in the same way as in the case of variations.

The advantage of giving measured items in the bills even though the quantities are not accurate, rather than allowing a provisional sum instead, is first, that unit rates are then available for use in valuing the work as remeasured. Secondly, the contractor is given a much more definite picture of the scope and nature of the work than he would otherwise have.

Claims

Claims for additional payments which have been accepted under clauses 4.23 and 3.24 of the JCT Form should each be fully and separately set out, detailing the calculations of the agreed sums. If the section is large enough to warrant it, a summary of the claims should be presented on its first page.

Fluctuations in costs of labour and materials, etc.

This is a particular form of claim arising under Fluctuations Option A, B or C of the JCT Form. The details should be fully set out to show how the amount of the claim has been calculated (see Chapter 8).

Measured contracts (that is, those based upon bills of approximate quantities or schedules of rates)

The final account for contracts, which require the complete measurement of the Works, will usually be a simpler, though more lengthy document. As there is no contract sum, the account will not be an adjustment account (i.e., showing deductions and additions), but will consist wholly of additions amounts.

Clause 4.3 of the JCT Form for use with bills of approximate quantities lists the components of the 'Ascertained Final Sum' and they may be conveniently grouped together, as follows. The same groupings may also be applied to schedule of rates contracts, as appropriate.

Summary of the account

The summary cannot commence with a contract sum as there is none, nor in consequence can it show any omissions. Instead, it will give a list of the succeeding sections with a total sum against each.

Bills of remeasurement

These will be similar in format to the measured sections of bills of quantities prepared for tendering purposes.

Preliminaries

This section will consist of the appropriate preliminaries items, valued in accordance with the Conditions of the contract.

Statutory bodies' accounts

This section will be similar to the 'adjustment of provisional sums' section of the account for a lump sum contract, but without the omissions amounts.

Claims

This section will be similar to that in an account for a lump sum contract (see the section 'Claims', p. 238).

Fluctuations

This section will also be similar to that in an account for a lump sum contract (see Section 'Fluctuations in costs of labour and materials', p. 239).

Prime cost contracts

The final account for prime cost contracts is in essence a very simple document. It shows the total of each of the constituents of the prime cost, i.e., labour, plant and materials and, in addition, either the fixed fee or the calculation of the fee where it is a percentage of the prime cost. If there is a target cost (see pp. 26–27), the resulting addition or deduction should also be shown. In the event of a situation arising being such as to warrant an adjustment of the fee, details of the calculation of the adjustment should also be given.

Final certificate

When the final account has been agreed, a photocopy of the original document (or, if sufficient, just the summary pages) will be sent to the architect or the Employer (if appropriate) along with a notification of the balance due to the contractor in settlement, after deduction of the total amount of interim certificates and any advance payments. It may be necessary to prepare a separate statement to show the reduced sum due, if the Employer has made payments when the contractor has defaulted. An example is where the contractor has failed to pay insurance premiums (clauses 6.4.3 and Insurance Option A.2 of the JCT Form). It should be noted that the Form does not require such amounts to be deducted from the contract sum, i.e., in the final account, but are to be deducted by the Employer from 'any sums due or to become due to the contractor'.

The architect is required to issue the Final Certificate within two months after the occurrence of the last of the events listed in clause 4.15, i.e.,

1 the end of the rectification period; where the Work has been broken into sections and there is more than one rectification period, then this criterion would apply to the last period to expire;
2 the date of issue of the Certificate of Making Good, or the last such certificate where the Work is carried out in sections;
3 the date when the architect sends the contractor copies of any outstanding ascertained loss and/or expense calculations and the final adjustment of the contract sum.

The Final Certificate should state (a) the adjusted contract sum or the ascertained final sum (i.e., the total of the final account), (b) the total of amounts paid on account, (c) any advance payment, and (d) the difference between the amounts expressed as a balance due from one party to the other. Such balance is stated to be 'subject to any deductions authorized by the Conditions', such as those instanced in the preceding paragraph. A further requirement of the Final Certificate is that it should state the basis on which the amount has been calculated, e.g., reference could be made to the appropriate contract conditions that determine the adjustments that have been made as well as the final account documentation previously supplied to the contractor.

Payment

To comply with the Housing Grants, Construction and Regeneration Act the Employer, within 5 days of issue of the Final Certificate, is to give the contractor a written notice of the balance, if any, due to the contractor. Furthermore the Employer is to inform the contractor the amount he proposes to pay, what the payment relates to and the basis for its calculation (see clause 4.15.3). As explained in Chapter 10 the JCT had to put this clause into the contract to comply with the requirements of the HGCRA. It is thought that most Employers will not bother to provide this written notice where they accept the amount shown in the Final Certificate (see clause 4.15.5). However, failure to provide this initial written notice will not prevent an Employer from subsequently paying the contractor a lesser amount than that stated in the Final Certificate. The Employer has 28 days from date of issue of the Final Certificate to pay any sums due to the contractor. If the Employer wishes to withhold or deduct monies from the sum due, he must give the contractor a written notice at least 5 days before the final date for payment, i.e., no later than 23 days from the date of issue of the certificate. In the written notice the Employer must state the amount(s) he intends to withhold and the ground(s) for withholding the money. If the Employer fails to provide this notice then he is legally obliged to pay the amount stated in the Final Certificate. Therefore, if there are any outstanding monies due to the Employer, e.g., claims for liquidated damages, the Employer must be advised about the need to issue this withholding notice, as there will obviously be no more payment certificates issued after the Final Certificate. This is the Employer's last opportunity to withhold payment from monies that are due to the contractor.

Liquidated and ascertained damages

Clause 2.32 of the JCT Form provides for liquidated damages to be paid or allowed by the contractor to the Employer calculated at the rate stated in the Contract Particulars, when a contract exceeds the contract period or, if an extension has been granted the extended contract period. The Employer's right to claim liquidated damages is conditional upon the architect issuing a Non-Completion Certificate for the Works or a section of the works where there is to be sectional completion, and the Employer issuing a notice as specified in clause 2.31.2.

Clause 2.32.2.2 says that the Employer may deduct any sum for liquidated damages from 'monies due to the contractor'. As far as the surveyor is concerned, therefore, the deduction of any amount for liquidated damages due should be made by the Employer from the sum stated in a certificate issued by the architect, as due to the contractor. All that the surveyor should do is to notify the architect of the amount of liquidated damages payable, so that the architect may in turn inform the Employer of the amount which he is entitled to deduct, if he so wishes, and remind the Employer of the need for a withholding notice. The surveyor should not deduct liquidated damages in the final account.

Effect of the final certificate

A number of legal and administrative misconceptions sometimes exist following the issue of the Final Certificate. Clause 1.10 of the JCT Form sets out the current position regarding the issue of the Final Certificate to the effect that it is conclusive evidence:

1 That where the quality of goods, materials or workmanship have been expressly described in the contract documents as having to be to the architect's approval – then the quality of such items have been executed to the architect's reasonable satisfaction. It is important to understand the severe limitations of this clause, i.e., it only applies to an item that is specified as having to be *to the architect's approval*, a situation that would rarely arise in contract documentation. Where the quality and standards of goods, materials and workmanship has been fully specified in the contract documentation, then the issue of the Final Certificate is not conclusive evidence that this quality has been met. As a result a contractor would still be liable to the Employer, after the issue of the Final Certificate, where it is discovered that materials and/or workmanship have failed to meet this specified standard.
2 That all contract terms have been complied with that require any addition, deduction or adjustment of the contract sum and that the contract sum has been correctly adjusted. However this clause is then heavily qualified to allow for later adjustments as a result of accidental inclusions or exclusions and arithmetical mistakes.
3 That all extensions of time due under clause 2.28 have been properly dealt with.
4 That the reimbursement of any direct loss and/or expense payable to the contractor under clause 4.23 (and which has been incorporated into the contract sum) is in final settlement of all and any claims that the contractor may have.

If any of the above events is in dispute and has been referred to adjudication, arbitration or other proceedings before the issue of the Final Certificate, then the issue of the certificate will have no effect on these disputed areas until

1 the proceedings have been concluded, whereupon the final certificate will be effective subject to the award or agreement reached, or
2 12 months has elapsed since the issue of the final certificate and neither party has taken further steps to commence proceedings. In this case the final certificate becomes effective, subject to any agreement that the parties may have reached during this period.

Similarly, either party may still refer a dispute to adjudication, arbitration or other proceedings within 28 days of issue of the final certificate, and as a result the final certificate will not be effective in respect of those matters that have been referred. It can be seen that either party has a relatively short period of time to consider the implications of the issue of the final certificate and failure to raise a dispute within 28 days could lead to a considerable loss (*Cambs. Construction Ltd v Nottingham Consultants (1996) 13-CLD-03-19*).

Example of final account

Figure 11.1 shows a method of setting out the summary of the final account for the contract detailed in Appendix A.

Reference

1. *Standard Method of Measurement of Building Works, Seventh Edition* (London: RICS Books, 1988), p. 14.

12

Cost control

The surveyor's role

Cost control has been defined as 'the controlling measures necessary to ensure that the authorized cost of the project is not exceeded'.[1] Initially, the 'authorized cost' is the contract sum, but at subsequent times in the construction period it will need to be adjusted to take account of necessary savings and of additional costs, which the Employer is willing and able to meet.

The 'controlling measures' are usually in the hands of the Employer and his architect. The surveyor is not normally in a position to exercise control of cost himself unless, as is sometimes the case, he has been appointed project manager, when he has overall control of the project. His usual role, however, is that of adviser to those who have that control and so is one of monitoring and reporting on costs rather than controlling them.

The contractor's surveyor likewise is unlikely to have any direct control of the costs of construction but is in the position of monitoring costs and expenditure and advising the contracts manager on such matters as the value of work done to date and the likely eventual total value of the contract. The contracts manager will be concerned with the relationship between the contractor's expenditure on the construction work and the total of payments received or anticipated from the Employer. While the payments are in excess of the expenditure, there is a profit. If in any part of the work, however, the contractor's costs exceed the amounts payable under the contract terms, there will exist a loss situation, at least so far as that section of the work is concerned. The contractor's surveyor will be expected to be able to explain why this situation has arisen and may even be expected to foresee a potential loss-making position developing and to warn the contracts manager so that he may take what steps are necessary to minimize or reverse the loss.

Such a position may arise from any one or more of the following causes

(a) inefficient deployment of resources (labour, plant and materials);
(b) excessive wastage or theft of materials;
(c) plant being allowed to stand idle or under-utilized;
(d) adverse weather or working conditions;
(e) delays arising from one or more of a variety of causes;
(f) under-pricing of tender documents by assumptions in regard to labour times, types and sizes of plant, etc., which do not equate with the realities of the construction work.

Recovery of additional expenditure under some of the foregoing heads may be possible (e.g., items (c) and (e) on the list) by successful claims within the provisions of the contract, but this will not be so in all cases.

Cost control from the viewpoint of the contractor's surveyor, therefore, is distinct from that of the surveyor. Although the concern of both of them centres around the total value of a project, the former is concerned with the relationship between the contractor's expenditure and payments to him by the Employer, while the latter is concerned with the relationship between the total amount that the Employer is willing and able to spend and the total sum that he is liable to pay under the terms of the contract.

In order to provide a basis for controlling cost, there must be a contract sum or an estimated value of the initial contract. In the case of lump sum contracts, there will be a contract sum, but in the case of prime cost and schedule of rates contracts there will not, and it will be necessary for the surveyor to prepare an estimate of contract value, preferably in the form of a cost plan. This will provide the basis for the necessary control mechanism, if cost is to be controlled at all.

Monitoring and reporting on the financial position

The Employer will need to be kept informed by the surveyor of the financial position of the contract. This is because he will be able to make the necessary arrangements to have the finance available to pay the contractor within the stipulated time upon presentation of interim certificates and also because he may be warned of any possibility or likelihood of the 'authorized cost' of the project being exceeded. He may then take steps, as considered appropriate, to increase his borrowing capacity or to effect savings on the project to ensure that the final cost is within his budget limit.

There are thus two aspects to financial reports (a) the current position and (b) the likely total cost. In order to carry out his duties adequately, the surveyor will have to take account of all the actual or likely causes of adjustment to the contract sum (as far as he is able to do so), as follows:

(a) expenditure of provisional sums;
(b) variations;
(c) claims;
(d) recoverable fluctuations in costs.

Expenditure of provisional sums

Where provisional sums are to cover the cost of such matters as telephone calls made on behalf of the Employer or daywork, then estimates of the likely expenditure will have to be included if actual costs are not known. Where provisional sums are to cover work to be carried out by the general contractor, the actual cost of the work (if already carried out and measured and valued) should be substituted for the respective provisional sums. If the work has not yet been measured and valued, then an estimate of the likely cost should be made and substituted for the provisional sums. Such estimates will usually be done by the 'approximate quantities' method, i.e., by measuring the main items of work involved reasonably accurately (but not necessarily exactly). The measured items are then priced at 'all-in' rates, i.e., at rates (based on bill rates where possible) which allow not only for all the work described in each item but also for the cost of all related items which have not been separately measured.

Variations

Where variations have been included in Architect's Instructions and have been accurately measured and valued, the net extra cost or saving (i.e., taking both additions and omissions into account) should be incorporated in the financial assessment. In regard to those variations which have not been measured and valued, including any which are expected but have not yet been ordered by the architect, the total net value of each should be estimated in the manner described under 'Expenditure of provisional sums' and be taken into account.

Claims

If claims can be dealt promptly, the amounts allowed can be taken into account when the next financial statement is made. Claims, which have been presented but are still outstanding, should be valued as accurately as reasonably possible (erring, if anything, on the generous side) and included in the statement. If any claim is subsequently rejected, it will, of course, be omitted from the next statement. Any claims unlikely to be accepted can probably be safely ignored.

Fluctuations

Fluctuations, which are recoverable under the terms of the contract should be valued at least monthly for inclusion in interim valuations. Where the Formula Method is used, fluctuations will be calculated at each valuation date (see p. 164).

The amounts so agreed will be included in financial reports and an estimate of future payments will need to be made for the purpose of forecasting the likely total cost of the project.

It may be thought that interim valuations are, in effect, reports on the current financial position of a project. While to a large extent this is so, they do not necessarily give a complete picture. They give the accurate current position of expenditure by the Employer, but they do not necessarily include all known costs for which he is currently liable. The position is also clouded to some extent by the retention, particularly where there have been some releases, and also by the inclusion in valuations of the value of unfixed materials. Examples are given below of expenditure to which the Employer is committed but which may not be included in the latest interim valuation, often because of time lag. The value of unfixed materials, included in interim valuations, would be excluded from financial reports.

(a) Work which is covered by provisional sums or which is the subject of variation orders, which is still in progress but was insufficiently advanced for inclusion in the last valuation
(b) Fluctuations for which the Employer is liable but for which supporting details are awaited from the contractor
(c) Claims, which are expected to be accepted but not yet been settled.

The contractor's surveyor will be involved in preparing similar financial reports for the contracts manager, contracts director or managing director. His reports will, of course, be similar in most respects to those prepared by the Employer's surveyor. They may differ, however, in the following matters.

(a) Variations may be included which the contractor claims have been ordered but which have not been confirmed in writing. These may or may not be confirmed later (see pp. 112–113).
(b) The value of all claims will be included, whether accepted or not. A distinction may be shown, however, between those agreed or likely to be agreed and those unlikely to be accepted.
(c) The valuation of variations, remeasured work, claims and fluctuations will differ, partly because of the item rates and prices which remain to be agreed and partly because aspects of claims and fluctuations payments may be the subject of dispute.

Examples of financial reports prepared for the Employer are given in Figures 12.1 and 12.2.

As the effective control of cost lies with the architect, it is desirable that the surveyor should bring his influence to bear to encourage the exercise of that control. He can do so by providing cost reports on the current and future financial positions, which are prompt, reliable and up-to-date. It is also desirable that, before issuing instructions that affect the cost of the contract, the architect should consult the surveyor on the likely effect and obtain advice on the cost of any alternative that may be available.

Shops & Flats, 28-34 Thames St., Skinton

Financial Report No. 10

Current Position as at 26 September 2006

	£
Value of measured work as bills of quantities as at 14 September 2006	790,534
Value of preliminaries as at 14 September 2006	89,383
Payments to Statutory Bodies as Valuation No. 12	14,000
Variations	12,257
Fluctuations in labour and materials costs	2,364
Claim under clause 4.24, as agreed	1,610
Claim under clause 4.25 outstanding but for which liable	10,000
Value of work carried out to date	920,148
Total actual expenditure as Certificate No. 12 (before deduction of retention and advance payment)	913,962
Total anticipated expenditure to date as forecast*	977,191

29 September 2006

*see p. 257

Figure 12.1 Example of report on current financial position of a contract

Forecasting client's capital expenditure flow

If a contract is large with a cost of millions of pounds to be expended over several years, it is essential to make a reasonably reliable forecast of the likely flow of expenditure. Even where costs are more modest, and contract time correspondingly shorter, it is to a client's advantage to be able to anticipate his cash flow requirement and to be able to arrange for finance to be available accordingly.

In consequence, it is often part of the surveyor's task to prepare, either in tabular or graphical form, an anticipated 'rate of spend' or 'expenditure flow' forecast.

In order to do this, it is necessary to prepare a programme for carrying out of the various parts of the construction work.

Shops & Flats, 28-34Thames St., Skinton

Financial Report No. 11

Estimated total cost

	£
Contract sum	1,448,000
Less Contingency sum	15,000
	1,433,000
Less provisional sums (net omission)	1,200
	1,431,800
Variations	19,000
Fluctuations in labour and materials costs	5,000
Claims	11,610
	1,467,410
Contingency sum for remainder of contract	6,600
Estimated total cost of contract	1,474,010

29 September 2006

Figure 12.2 Example of report on likely total cost of a contract

A progress chart or 'bar chart', as illustrated in Figure 12.3, is commonly used to show such a programme. If the contractor has produced a programme, which he has submitted to the architect or project manager, or if he is willing to make his programme available to the surveyor, so much the better. Using bill rates and prices (or schedule rates and prices, if there is no bill), each section of the work is valued, the totals being broken down into monthly amounts over the period allocated to each section, as shown in Figure 12.3. The monthly amounts are then totalled and to each total is added the appropriate proportion of the Preliminaries. The final totals for each month (which together should equal the contract sum) can then be tabulated or plotted on a graph.

A graph plotted from such data is usually referred to as an 'S-curve' graph because of the shape of the graph. It indicates a

Contract programme										
Month	1	2	3	4	5	6	7	8	9	10
Operation										
Demolition	3,661	2,000								
Groundwork	15,820	23,264	29,964	20,100	18,983	40,014	37,967			
Mains services connection		2,000			3,000			3,000		
In situ concrete	2,833	12,655	16,054	16,054	6,044	13,221	15,677	18,886	20,210	16,999
Masonry			6,385	25,535	20,428	19,311	18,992	22,343		
Structural carcassing										
Metal/timber				6,871	12,168	14,243	15,603	11,380	11,309	
Cladding/Covering								8,693	8,202	9,415
Waterproofing		6,500	12,580					13,500	24,350	8,680
Windows/ Doors/Stairs							16,670	20,262	20,903	18,980
Surface finishes					5,422	4,831		3,543	4,563	4,455
Furniture/ Equipment								1,740		
Building fabric sundries						6,233	6,758	7,135	7,217	5,462
Disposal systems										10,855
Piped supply systems						1,228	2,456	2,456	2,305	
Mechanical heating			3,000		5,000	2,000	2,000	2,000	4,000	
Security								10,000		
Electrical services			2,000			2,000	2,000	2,000	2,000	2,000
Lift installations					32,910					
Drainage	941	1,053			2,456	2,536	1,962	1,675	1,675	1,866
Fencing					3,675	1,830	1,830	2,290	1,043	
Pavings										
Provisional sums			50	50	50	50	50	50	50	50
Preliminaries	10,417	10,350	9,350	6,350	6,350	6,350	6,350	6,350	6,350	6,350
Dayworks/ contingencies	1,062	1,061	1,062	1,061	1,062	1,061	1,062	1,061	1,062	1,061
Totals	34,734	58,883	80,445	76,021	117,548	114,908	129,377	138,364	115,239	86,173

Notes: Adjustments for errors in bills of quantities distributed over monthly values. Dayworks and contingencies have been included so that the total agrees with the Contact Sum (* relating to contract in Appendix A).

Figure 12.3 Use of programme chart for expenditure forecasting.

chart (simplified)*

11	12	13	14	15	16	17	18	19	20	Totals
										5,661
										186,112
	6,000			3,000			5,000	1,000		23,000
14,355	17,377	18,510								188,875
	12,608	27,291	6,703							159,596
										71,574
6,495										32,805
										65,610
18,082	19,108	14,235								128,240
1,127	4,563	4,563	2,416			5,153	5,153	4,133	3,758	53,680
				3,480	3,480	3,480	1,740			13,920
										32,805
10,855	8,112									29,822
	2,197	2,563	2,563	625		1,874	2,563	710		21,540
24,000			12,000	4,000	20,000	3,000	3,000	2,194		86,194
5,000					5,000				2,850	22,850
6,000	3,000			3,000	3,000		5,805			32,805
10,000	30,000									72,910
925	861									15,950
							4,580	1,845		17,093
	1,500	1,000	1,500				1,000	1,955	1,511	8,466
50	50	50	50	50	50	50	2,000	1,558	2,150	6,458
7,350	8,350	6,243	6,344	6,350	6,350	6,350	8,350	9,350	11,200	150,804
1,062	1,061	1,062	1,061	1,062	1,061	1,062	1,061	1,062	1,061	21,230
105,301	114,787	75,517	32,637	21,567	38,941	20,969	40,252	23,807	22,530	1,448,000

build-up in the rate of expenditure over the first two or three months, an acceleration of the rate over the main part of the contract period and a gradual run-down over the final three months or so. Figure 12.4 shows a typical S-curve graph for a £500,000 contract over a 15-month contract period.

A graph produced as described above will, of course, indicate anticipated cumulative monthly total values of work done and not a forecast rate of client's expenditure. The difference between the two is the amount of retention deductible during the contract period and the amounts of releases of retention at its end. From the amounts of monthly values of work the proportion of value to be retained should be deducted, and amounts of releases should be added. The latter may be only at the end of the contract period or may be earlier in the case of sectional completion. The adjusted amounts may then be plotted to give an S-curve showing expected rate of expenditure by the client.

There is an alternative method of producing such a graph. If the totals of interim valuations for a number of contracts are plotted and curves of 'best fit' are drawn, an S-curve of approximately the same shape will result for each contract. It follows, therefore, that

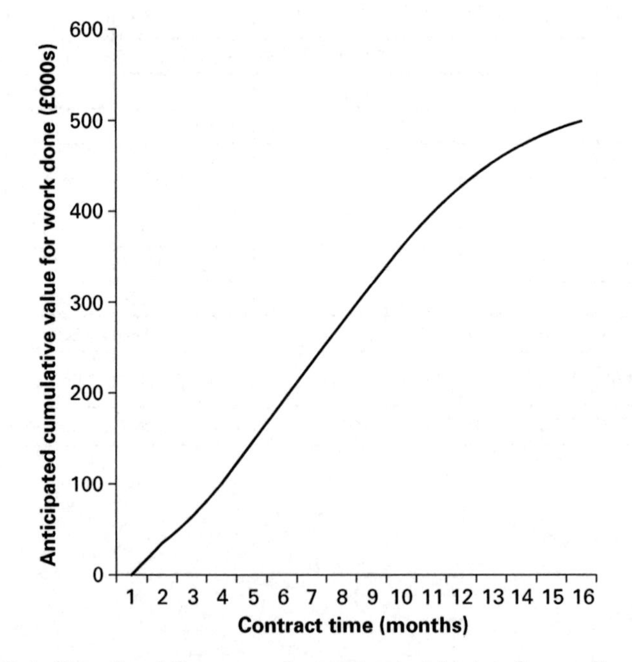

Figure 12.4 Standard S-curve of anticipated 'rate of spend'

it is possible, within a given range of cost and over a limited range of contract times, to produce a 'standard' S-curve, which can be used to predict expenditure flows for further contracts without the need to go through the detailed procedure described above.

A third method of producing an S-curve graph of value of work executed is by the use of a formula. This method was used by the Department of Health and by the Property Services Agency (now defunct).[2] It is really a development from the method described in the last paragraph because, having produced standard S-curves in the manner described, the Department of Health then proceeded to determine the underlying mathematical formula. The formula devised and used by the department is

$$y = S\left[x + Cx^2 - Cx - \frac{1}{K}(6x^3 - 9x^2 + 3x)\right]$$

where y = cumulative value of work executed in £
S = contract sum in £
x = proportion of contract time completed
C and K are parameters relating to contract value:

Contract value £s	*C*	*K*
10,000–50,000	−0.439	5.464
50,000–100,000	−0.370	4.880
100,000–200,000	−0.295	4.360
200,000–500,000	−0.220	3.941
500,000–2,000,000	−0.145	3.595
2,000,000–3,500,000	0.010	4.000
3,500,000–5,500,000	0.110	3.980
5,500,000–7,500,000	0.159	3.780
7,500,000–9,000,000	0.056	3.323
Over 9,000,000	−0.028	3.090

It will be obvious that the formula has to be used for each month's anticipated value of work done – which means 15 times for a 15 month contract, etc. The mathematical work entailed presents no great problem when a computer is available, but without one the volume and tedium of the necessary calculations may deter many surveyors from using this method.

It is important, of course, that having produced a forecast of rate of expenditure, the information should be given to the client, either direct or through the architect, so that he may use it to arrange finance for the project. It may be necessary on large jobs to revise the forecast rate of expenditure at any point during the contract period for the remaining period, in the light of actual expenditure, and if this is done, again, the client should be given the revised information without delay.

A further use of the S-curve graph is to plot the totals of interim valuations as they are done, thus producing a curve of actual expenditure. If the latter is seen to be beginning to fall short of the anticipated expenditure, it will provide early warning of possible trouble. There may be a satisfactory explanation, such as a prolonged period of bad weather or a strike in the construction industry or in

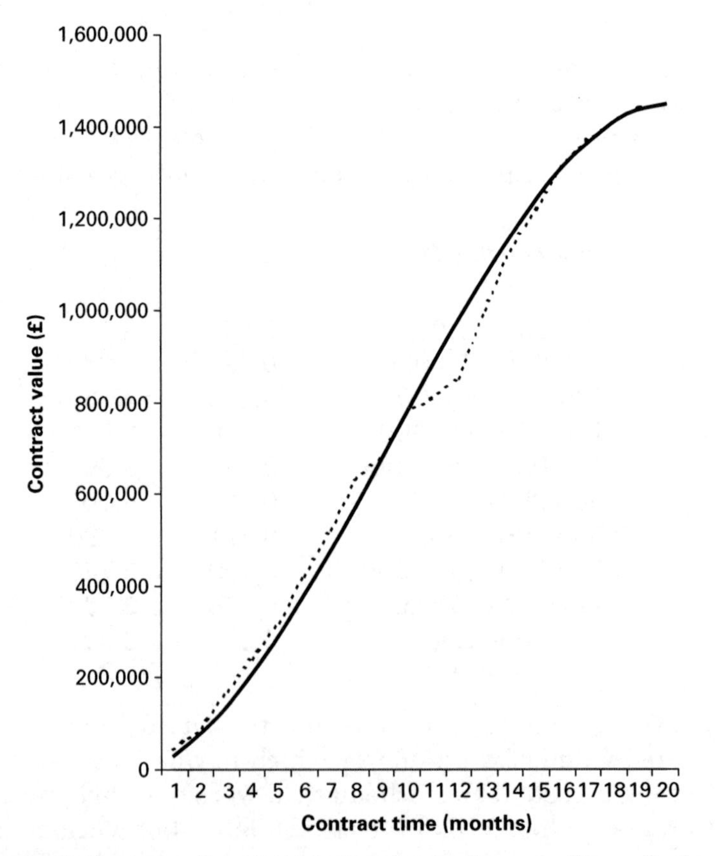

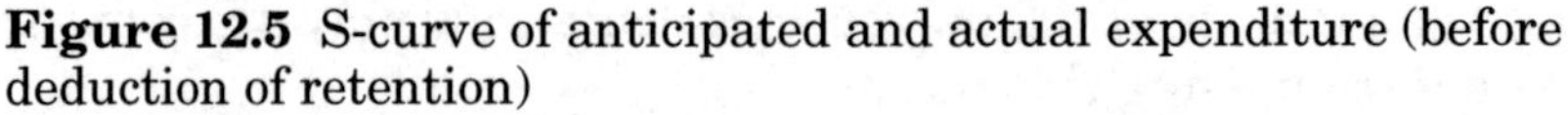

Figure 12.5 S-curve of anticipated and actual expenditure (before deduction of retention)

a related industry. If there is no obvious explanation, there may be an indication that the contractor is running into difficulties, financial or otherwise. The early warning may provide an opportunity to remedy the situation.

In such circumstances as those described in the last paragraph, the formula can be used (a) to determine a revised completion date; (b) to forecast likely expenditure in future months, based on actual expenditure in past months; and (c) to calculate the rate of expenditure necessary to achieve completion by the original date stated in the contract[3] (although experience has shown that this is seldom possible).

Figure 12.5 shows a graph of anticipated and actual expenditure for the contract of which particulars is given in Appendix A. All values shown are those before deduction of retention. Anticipated expenditure has been calculated using the DHS formula, using values $C = -0.145$ and $K = +3.595$. The calculated forecast of cumulative expenditure for each month is as follows:

Month	*Expenditure*	*Month*	*Expenditure*
1	£30,716	11	£878,272
2	£76,696	12	£977,191
3	£136,125	13	£1,071,435
4	£207,192	14	£1,159,193
5	£288,085	15	£1,238,650
6	£376,991	16	£1,307,995
7	£472,096	17	£1,365,415
8	£571,590	18	£1,409,097
9	£673,659	19	£1,437,230
10	£776,490	20	£1,448,000

The curve of actual expenditure shows monthly values before deduction of retention and exclusive of materials on site, fluctuations and claims, thus allowing an undistorted comparison to be made. It will be seen that the expenditure during the first nine months was higher than forecast, but after that it dropped markedly below the forecast level. The reasons for this were bad weather, labour shortages experienced by the contractor and the delay of work on the lift installations (see p. 228). From September 2006 onwards, the rate of production increased and by January 2007, actual expenditure had almost caught up with the forecast value.

By plotting actual expenditure each month, it was possible to see at the June 2006 valuation that the curve was flattening out, thus giving warning of possible problems ahead. In fact, the causes were identified, as noted above, and steps were taken to remedy the situation so that the contract was eventually completed on time.

Forecasting contractor's earnings and expenditure

The same information as is produced by the forecasting techniques described above is of value to the contractor. It enables him to monitor the financial progress of each of the contracts running at any time. The S-curve of anticipated cumulative values of work done represents to the contractor his anticipated or 'programmed' earnings. The rate of client's expenditure becomes, from his point of view, actual earnings.[4] A comparison between the two provides an indication whether the financial progress of the contract is satisfactory or not.

Contractor's cost control

Cost control is an essential part of any commercial organization. Unfortunately construction firms and associated sub-contracting organizations tend to have far greater problems in maintaining effective financial control systems than many other industries. Some of the reasons for these difficulties are considered below

1 labour does not work in a controlled environment (subject to weather conditions);
2 difficult to monitor labour on a large site;
3 difficult to secure materials on site (damage, vandalism, theft and waste);
4 working under continually changing conditions (different sites and location);
5 difficult to forecast demand, i.e., availability and cost of labour and materials;
6 alteration to work and design by the client (variations);
7 subject to interference and delay by other parties involved in the project (architect, clerk of works, statutory undertakers, building control officer).

Despite the above problems it is essential that contract costs are monitored. Construction firms traditionally operate on small profit margins and limited working capital, and as a result it is vitally important to try and maintain a positive cash flow. Two ways in which a contract's profitability may be reviewed are through the use of cost/value reconciliations and site costings.

Cost/value reconciliation

The Employer regularly finances construction contracts through interim valuations. If the interim payment is an accurate reflection of the works carried out, then a comparison may be made between income and expenditure and so provide an interim report of the contract's profitability. This monitoring process is based upon a comparison between the monies received from the client via interim certificates and the costs incurred by the contractor in executing the work. However, this process is not always straightforward, as the date of interim valuations may not coincide with the normal end-of-month accounting procedure used by a contractor. There may also be areas of dispute within the interim valuation with regard to the payment of preliminary items, variations, claims, defective work as well as areas of work that may have been accidentally or deliberately undervalued/overvalued; finally there may be similar problems occurring between the contractor and sub-contractors. These problems may not be resolved until the final account stage and therefore any profit calculations should only be viewed as being approximate. Figure 12.6 provides a simplified example of a cost/value reconciliation statement where it can be seen the contractor currently achieving a profit of 4.47%. Although the use of cost/value reconciliations should make it possible to identify any underlying trends, i.e., a steady fall in profitability, or a sudden change that would require immediate investigation will not identify in detail where a profit or loss is being made.

Standard costing

Where more detailed information is required it will be necessary to make use of site costing techniques. The advantage of such an approach is that a contractor can be provided with information that can be used far more effectively to control project costs. The

Beecon Ltd

Cost/value reconciliation

Shops and Flats, Skinton

Revenue			£
Interim Certificate No. 12			934,077.00
Adjustments for overclaims:			
In situ concrete work	−2,000.00		
Windows and doors	−5,000.00		
Materials on site	−600.00	−7,600.00	
Adjustment for extras			
Claim not yet agreed	10,000.00		
Variations still to be agreed	1,000.00		
Fluctuations not yet submitted	250.00	11,250.00	3,650.00
			937,727.00
Expenditure			
Site staff	28,800.00		
Site operatives	48,547.00		
Labour-only sub-contractors	98,542.00		
Payments to sub-contractors	470,590.00		
Materials as invoiced	203,654.00		
Materials delivered – not invoiced	500.00		
Site expenses	6,000.00	856,633.00	
Head office costs: 5% of expenditure		42,832.00	899,465.00
Difference			38,262.00
Difference as % of expenditure			4.47%

Figure 12.6 Example of a contractor's cost/value reconciliation

principle of site costing techniques is to monitor individual operations so that any deviations will be identified very quickly. One examples of this costing technique is standard costing.

Standard costing is based upon identifying the actual cost (in money or in hours) of an operation and comparing that cost with the estimator's target for that work. The basic principles of standard costing are

1 there must be a set standard or target for the operation – this would normally be available through the estimator's worksheets;

Beecon Ltd — Week No. 5

Cost statement: Shops and Flats, Skinton

M	T	W	T	F	S	S	Total Hours	Measure	Standard	Total	Operation	Hours Lost	Hours Gained
8	8						16	150 m^3	0.10	15.0	Basement excavation	1.0	
16	8						24	150 m^3	0.15	22.5	Ditto plant hire	1.5	
	4	12					16	84 m^2	0.30	12.6	Hardcore	3.4	
	4	6					10	84 m^2	0.15	12.6	Ditto plant hire		2.6
		8	32				40	375 m^2	0.10	37.5	Blinding	2.5	
		16	16	4			36	80 m	0.50	40.0	Formwork		4.0
				16	6		22	100 m^2	0.15	15.0	Mesh reinforcement	7.0	
	Total labour hours						130.0			120.1		13.9	4.0
	Total plant hours						34.0			35.1		1.5	2.6

Figure 12.7 Example of a contractor's cost statement

2 the contractor's surveyor must monitor actual performance – this may be achieved through site records or payment records;
3 the contractor's surveyor may then compare the actual performance with the set standard;
4 where there is a variation between actual and standard performance, the reason(s) for the variation should be investigated and identified;
5 the result of the investigation should be notified to the estimating department if the set standard appears to be at fault and/or corrective action should be implemented on site where possible.

The above information may be presented to the site management in the form of a standard cost statement (see Figure 12.7). From this statement it is possible to study the individual performance of various operations and to calculate from the standard cost what the contract's performance has been for that week, i.e., labour gains of 4 h and losses of 13.9 h, giving an over-all loss of 9.9 h.

The value of such comparisons, if carried out at least once a month, is that they enable the contractor to have early warning of a loss-making situation before the position becomes irretrievable. By pinpointing where the losses are being made, the contractor can take action to remedy the situation. Again, prompt and up-to date information is the key to practical use of financial analysis. Otherwise, its value is merely historical. Thus, the contractor's surveyor who keeps the contracts manager regularly and accurately informed on the current relationship between earnings and expenditure is of considerable value to the contractor.

References

1. AQUA GROUP, *Contract Administration for Architects and Quantity Surveyors, seventh Edition* (London: Collins, 1986), p. 14.
2. HUDSON, K. W., DHSS expenditure forecasting method, *Chartered Surveyor, BQS Quarterly*, Vol. 5, No. 3, Spring 1978, p. 42.
3. COST INTELLIGENCE SERVICE, DHSS, *Health Notice HN (78) 149 Forecasting Incidences of Expenditure, reference* CON/C462/17, November 1978, pp. 2–3.
4. GOBOURNE, J., *Cost Control in the Construction Industry* (London: Butterworth & Co. Ltd, 1973), pp. 7–8.

13

Capital allowances

Introduction

Clients' accountants regularly request information about the value of plant and machinery included in interim payments to contractors. Why do they require this information, and what is their interpretation of the terms plant and machinery?

This chapter examines the basic principles of capital allowances and considers the specific application of these principles in relation to construction and buildings.

Taxation is often considered to be a complex specialization and an area in which only taxation experts should be involved. While this is true it is essential for all surveyors to have a basic understanding of capital allowances in the context of construction and buildings. This basic understanding should be sufficient to enable them to advise and help clients or employers with regard to:

(i) preparing calculations to support claims for capital allowances;
(ii) understanding the use of tax planning to support effective design decisions.

This chapter identifies the importance of surveyors understanding capital allowances in order to ensure that client's interests are properly protected.

Capital allowances

Capital allowances are available to taxpayers against capital expenditure on certain types of fixed assets. Capital allowances reduce the tax liability of a business. There are a number of different categories of qualifying expenditure which can attract capital allowances. These categories include:

(i) Plant and machinery allowances
(ii) Enhanced capital allowances (ECA) (energy saving and environmentally beneficial plant and machinery)
(iii) Industrial buildings allowances
(iv) Hotel buildings allowances
(v) Buildings in enterprise zones
(vi) Agricultural buildings allowances
(vii) Flat conversion allowances
(viii) Mineral extraction allowances
(ix) Research and development allowances (formerly Scientific Research)
(x) Know-how allowances
(xi) Patents
(xii) Dredging
(xiii) Assured tenancy allowances.

The principal legislation governing capital allowances is the Capital Allowances Act 2001. There are several general requirements that must be satisfied if expenditure is to be eligible for capital allowances. First, it must be expenditure incurred by the person or organization claiming the allowance; second, the expenditure must have arisen in the course of a trade; third, the expenditure must have occurred in a chargeable period; and fourth, the asset must be owned by the person or organization claiming the allowance.

Expenditure can qualify for relief even if it is funded by a loan or hire purchase agreement. In both these situations, only the cost of the asset is allowable (although interest and hire purchase charges are allowed separately as a business expense). However, in the case of assets provided under a lease agreement, it is the lessor who is entitled to the capital allowances at this present time (leasing eligibility currently under review).

Capital allowances are available in respect of chargeable periods. Allowances are first given in the chargeable period in which expenditure is incurred; this is known as the basis period.

For a company, the chargeable period will be the accounting period.

The allowances most often available in respect of property are plant and machinery, energy saving and environmentally beneficial plant and machinery, industrial buildings, hotels and buildings in enterprise zones. These allowances will be considered in further detail.

Plant and machinery

One of the main difficulties in claiming allowances for plant and machinery stems from the problem of deciding which assets are included in each of the terms plant and machinery. There is no definition of these terms within the taxation statutes. 'Machinery' is reasonably straightforward; 'plant', however, is more complex, and, as a result, there is much case law concerning its definition. The key distinction between what can and cannot be claimed as plant is that apparatus used to carry on a business is 'plant', whereas the setting in which the business is carried on is not plant. Loose items of equipment do not generally cause problems in classification; it is those items of plant and machinery that are fixtures within, or form part of, a building or structure that give rise to difficulties.

Expenditure on buildings which does not qualify for allowances (CAA 2001, s21)

Subject to List C below, expenditure on the provision of a building will not qualify for plant and machinery allowances where it is incurred after 29 November 1993. For these purposes the expression 'building' includes:

(a) any assets incorporated in the building;
(b) any assets not incorporated in the building, because they are movable or for some other reason, but are nevertheless of a kind which are normally incorporated into buildings;
(c) any of the following:

(CAA 2001, s21, List A):

1 walls, floors, ceilings, doors, gates, shutters, windows and stairs;
2 mains services and systems, for water, electricity and gas;
3 waste disposal systems;
4 sewerage and drainage systems;

5 shafts or other structures in which lifts, hoists, escalators and moving walkways are installed;
6 fire safety systems.

It would appear that there is room for debate over the meaning of 'normally' in (b) above. What may be normal for one type of building may not be normal for another.

Expenditure on structures, assets and works not qualifying for capital allowances (CAA 2001, s22, List B).

For the purposes of these provisions the word 'structure' means a fixed structure of any kind, other than a building (CAA 2001, s22 (3) (a)). A structure is any substantial man-made asset (Inland Revenue Press Release 17 December 1993).

Subject to List C below, expenditure on the provision of a structure or other asset listed in (1)–(7) below, or on any works involving the alteration of land, will not qualify for plant and machinery allowances where it is incurred after 29 November 1993.
(CAA 2001, s22, List B):

1 A tunnel, bridge, viaduct, aqueduct, embankment or cutting.
2 A way or hand standing, such as a pavement, road, railway or tramway, a park for vehicles or containers, or an airstrip or runway.
3 An inland navigation, including a canal or basin or a navigable river.
4 A dam, reservoir or barrage including any sluices, gates, generators and other equipment associated with the dam, reservoir or barrage.
5 A dock, harbour, wharf, pier, marina or jetty, and any other structure in or at which vessels may be kept or merchandise or passengers may be shipped or unshipped.
6 A dike, sea wall, weir or drainage ditch.
7 Any structure not in (1)–(6) above, with the exception of an industrial structure (other than a building) which is or is to be an industrial building within the meaning of CAA 2001, Pt3, Ch2, a structure in use for the purposes of an undertaking for the extraction, production, processing or distribution of gas, and a structure in use for the purposes of a trade which consists in the provision of telecommunication, television or radio services.

For the purposes of these provisions, the alteration of land does not include the alteration of buildings or structures.

Expenditure on buildings not excluded from qualifying for allowances (CAA 2001, s23, List C)

Expenditure incurred on the provision of any assets listed in (1)–(33) below is not excluded by above from qualifying for plant and machinery allowances. An asset does not, however, fall within (1)–(16) below (and is thus excluded from qualifying as plant or machinery) if its principal purpose is to insulate or enclose the interior of the building or to provide an interior wall, a floor or a ceiling which (in each case) is intended to remain permanently in place.

(CAA 2001, s23, List C):

1 Machinery (including devices for providing motive power) not within (2)–(33) below.
2 Electrical systems (including lighting systems) and cold water, gas and sewerage systems provided mainly to meet the particular requirements of the qualifying activity, or provided mainly to service particular plant or machinery used for the purposes of that activity.
3 Space or water heating systems; powered systems of ventilation, air cooling or air purification; and any ceiling or floor comprised in such systems.
4 Manufacturing or processing equipment; storage equipment, including cold rooms; display equipment; and counters, checkouts and similar equipment.
5 Cookers, washing machines, dishwashers, refrigerators and similar equipment; washbasins, sinks, baths, showers, sanitaryware and similar equipment; and furniture and furnishings.
6 Lifts, hoist, escalators and moving walkways.
7 Sound insulation provided mainly to meet the particular requirements of the qualifying activity.
8 Computer, telecommunication and surveillance systems (including their wiring or other links).
9 Refrigeration or cooling equipment.
10 Fire alarm systems; sprinkler equipment and other equipment for extinguishing or containing fire.

11 Burglar alarm systems.
12 Strong rooms in bank or building society premises; safes.
13 Partition walls, where movable and intended to be moved in the course of the qualifying activity.
14 Decorative assets provided for the enjoyment of the public in the hotel, restaurant or similar trades.
15 Advertising hoardings; and signs, displays and similar assets.
16 Swimming pools (including diving boards, slides and structures on which such boards or slides are mounted).
17 Any glasshouse constructed so that the required environment (i.e., air, heat, light, irrigation and temperature) is controlled automatically by devices forming an integral part of its structure.
18 Cold stores.
19 Caravans provided mainly for holiday lettings.
20 Buildings provided for testing aircraft engines run within the buildings.
21 Movable buildings intended to be moved in the course of qualifying activity.
22 The alteration of land for the purpose only of installing plant or machinery.
23 The provision of dry docks.
24 The provision of any jetty or similar structure provided mainly to carry plant or machinery.
25 The provision of pipelines, or underground ducts or tunnels with a primary purpose of carrying utility conduits.
26 The provision of towers to support floodlights.
27 The provision of any reservoir incorporated into a water treatment works or the provision of any service reservoir of treated water for supply within any housing estate or other particular locality.
28 The provision of silos provided for temporary storage or the provision of storage tanks.
29 The provision of slurry pits or silage clamps.
30 The provision of fish tanks or fish ponds.
31 The provision of rails, sleepers and ballast for a railway or tramway.
32 The provision of structures and other assets for providing the setting for any ride at an amusement park or exhibition.
33 The provision of fixed zoo cages.

Expenditure unaffected by Lists A & B CAA 2001, s23

Expenditure incurred on any of the following do not fall within the scope of expenditure prohibited from qualifying for plant and machinery allowances and continues to be regulated by specific rules:

1 Thermal insulation within CAA 2001, s28
2 Fire safety within CAA 2001, s29
3 Safety at sports grounds within CAA 2001, ss30–32
4 Personal security assets within CAA 2001, s33
5 Films, tapes and discs dealt with in accordance with an election under ITT01A 2005, s143 or F (No 2) A 1992, s40D
6 Computer software within CAA 2001, s71.

The items that can be claimed, therefore, extend beyond those normally considered as plant by most surveyors. It should also be noted that consideration of available allowances should be taken into account at the design stage; for example, a plenum air-conditioning system may become a financially favourable alternative when the available capital allowances are taken into account in a comparative design study. When considering the installation of plant and machinery in an existing building where a trade is carried on, the work incidental to this installation may be treated as plant and machinery.

Legal cases concerning the definition of plant and machinery can help in calculating and negotiating claims for allowances and in using tax planning as part of the design process to maximize the benefits available. Some of the decisions defining plant are summarized below.

1. Wimpy International Ltd v Warland (1988); Associated Restaurants Ltd v Warland (1989) STC 273, 61 TC 51

Both taxpayers were members of the same group and operated restaurants under the names 'Wimpy' and 'Pizzaland'. Each refurbished their restaurants and claimed that various items should be treated as plant. The majority of the relevant items served, it was claimed, to attract custom and provide an atmosphere conducive to the enjoyment of meals. The claim in respect of most items was dismissed. Allowances were given in respect of special lighting which served a business purpose, namely to 'create an atmosphere of brightness and efficiency, suitable to the service and consumption of fast food meals and attractive to potential customers looking in from outside'. Also allowed, as plant was expenditure on one very special suspended ceiling also designed for purposes of 'atmosphere'. The items not allowed as plant included:

(a) shop fronts
(b) floor and wall tiles
(c) some suspended ceilings
(d) mezzanine floor
(e) trapdoor and ladder
(f) decorative brickwork.

2. *Leeds Permanent Building Society 1982 CD*
It was held that decorative screens which incorporated the organization's logo were plant and machinery.

3. *Hampton v Fortes Autogrill Ltd 1980 CD STC 80*
It was decided that a false ceiling installed as cladding for mechanical and electrical services installations was not plant and machinery.

4. *Jarrold v John Good & Sons Ltd (1962) 40 TC 681*
A company operating as a shipping agent and warehouse keeper, installed a quantity of movable partitions, which to all intents and purposes functioned as internal walls, albeit not load bearing. The partitioning consisted of metal ribs into which insets of either hardboard sheeting, doors or windows were installed. It was then screwed to the floor and ceiling to form a room of any desired size. The taxpayer claimed these partitions were necessary for the functioning of its trade, as departments could expand or contract, appear or disappear according to the volume of trade. At the planning stage, special instructions were given to the architects that the portion of the building to be devoted to offices was to be capable of the greatest degree of elasticity. A key observation was made by J. Pennycuick regarding the concept of setting: 'It appears to me that the setting in which a business is carried on, and the apparatus used for carrying on a business, are not always necessarily mutually exclusive.' The partitions were not excluded from being plant merely by virtue of the fact that they might be 'setting', and on the facts of the case, allowances were deemed to be available.

5. Dixon v Fitch's Garage Ltd 1975 CD STC 480
A canopy covering the service area of a petrol station was held not to be plant.

6. Hunt v Henry Quick Ltd; King v Bridisco Ltd (1992) STC 633
The facts in these two cases were practically identical and the appeals were heard together. Each taxpayer had erected a mezzanine platform in a warehouse in order to provide extra space for the storage of goods. These mezzanines comprised free-standing raised platforms on steel pillars with steel beams supporting wooden flooring. They were installed by a specialist firm, and it was admitted that it would have been difficult to buy an 'off the shelf' floor in the market place. The platform was bolted to the floor of the building only to ensure rigidity and safety, not for any reason of structural support. The company had applied for dispensation from building regulations, and in that application the platform was described as 'free-standing and not attached to the building in any way ... will be used exclusively for storage and no person will be resident on it'. The Commissioners had attached great weight to the lack of access by the public. On the grounds that the platforms were movable and temporary structures (they could be dismantled in three to four days if trading requirements changed) it was held that they did constitute plant for the purposes of capital allowances. However, lighting fixed to the underside of the platforms to illuminate the floor space below, did not qualify, following dicta in *Wimpy International v Warland*, as the lighting was not specific to the trade.

7. CIR v Scottish Newcastle Breweries Ltd (1982) STC 296, 55 TC 252
The taxpayer company operated a chain of hotels and licensed premises, either purpose-built or purchased as a shell and fitted out to the company's specification. It was claimed that one element of the company's trade was the provision of a certain type of atmosphere, or ambience, conducive to attracting custom. This argument proved attractive to the courts, and consequently allowances were given in respect of items used to create this 'ambience' such as décor, murals and sculptures. In the Court of Session, Lord Cameron thought that the terms 'plant' and 'setting' could overlap, and also stated:

> ... the question of what is properly to be regarded as 'plant' can only be answered in the context of the particular industry concerned

and, possibly in light also of the particular circumstances of the individual taxpayer's own trade.

Similar thoughts were expressed by Lord Wilberforce in the House of Lords: 'In the end each case must be resolved in my opinion by considering carefully the nature of the particular trade being carried on.'

8. Cole Bros Ltd v Phillips (1982) STC 307, 55 TC 188
John Lewis Properties Limited incurred expenditure on an entire electrical system installed in a store at the Brent Cross shopping centre, leased to Cole Bros Ltd. It was accepted that certain items were plant. These included:

- wiring to heaters, alarms, clocks, cash register;
- telephone trunking;
- wiring to lifts and escalators;
- emergency lighting system;
- standby supply system.

Other items were disputed by the Revenue, being mainly lighting and wiring, some of it specially designed, and also transformers and switch gear.

The case eventually reached the House of Lords. It was held that the 'multiplicity of elements' in the installation and the differing purposes to which they were put precluded one from regarding the entire electrical installation as a single entity. The claim therefore had to be approached on a piecemeal basis. The transformers and switch gear were held to be plant; other items were not, being a necessary part of the building or 'setting'.

Capital allowances for plant and machinery are available on a reducing balance basis. The allowance is 25% per annum. Generally, expenditure on plant and machinery is grouped together into a single pool; however, there are some exceptions. If assets are sold or disposed of, the sale price or market value (assume this is less than the original purchase price) is compared with the unrelieved expenditure. Where the sale price exceeds the unrelieved expenditure, the excess is added to taxable income as a balancing charge. A balancing allowance is available for the shortfall between sale price and unrelieved expenditure.

Example 1

A company has the following plant and machinery transactions in the accounting periods ending 31 December 2004 and 2005.

	£
April 2004 Purchase of plant	50,000
November 2004 Sale proceeds	5,000
February 2005 Purchase of plant	5,000
March 2005 Purchase of plant	110,000
June 2005 Sale proceeds	25,000
September 2005 Purchase of plant	16,000

The pool balance brought forward to 1 January 2004 is £55,000. The allowances are as follows:

For the accounting period ending 31 December 2004

	£
Pool value brought forward	55,000
Additions	50,000
	105,000
Less sales proceeds	(5,000)
	100,000
Writing down allowance @ 25%	(25,000)
Pool balance carried forward to 2005	75,000

For the accounting period ending 31 December 2005

	£
Pool value brought forward	75,000
Additions	131,000
	206,000
Less sales proceeds	(25,000)
	181,000
Writing down allowance @ 25%	(45,250)
Pool balance carried forward to 2006	135,750

Taxable profit for the accounting periods to 31 December 2004 and 2005 is therefore reduced by £25,000 and £45,250 respectively due to the available capital allowances for plant and machinery.

Different building uses will obviously attract different levels of allowance for plant and machinery. Figure 13.1 shows how capital allowances for plant and machinery could arise in an office.

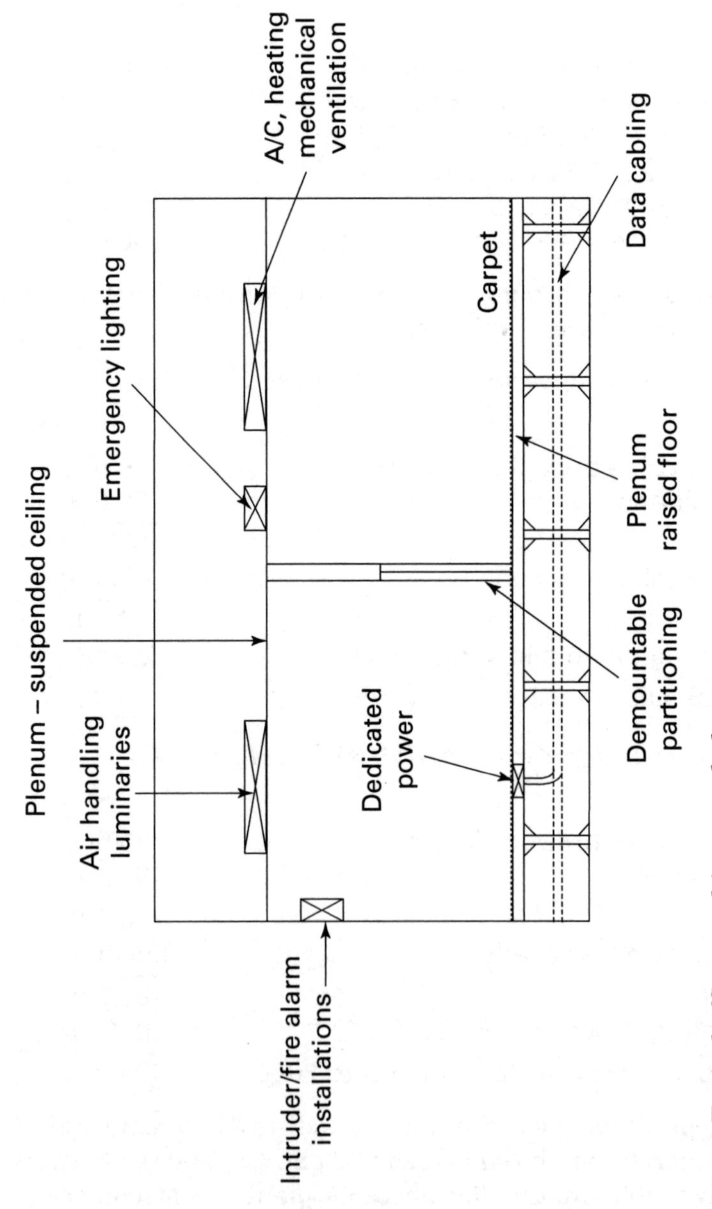

Figure 13.1 Typical office: machinery and plant

Enhanced capital allowances – energy saving and environmentally – beneficial plant and machinery

The Finance Act 2001 introduced a 100% first-year allowance on energy-saving plant or machinery. The introduction of this allowance is part of a wider package of measures aimed at helping businesses to reduce their energy consumption and so help the United Kingdom reduce emissions of greenhouse gases.

The expenditure must be on plant that is new (i.e., not second-hand), and must be incurred on or after 1 April 2001.

In order to qualify, plant must be of a description specified by Treasury Order, or must meet the energy-saving criteria specified by Treasury Order for plant of that description. The Treasury may also require that a certificate of energy efficiency is in force, issued by the Secretary of State or devolved equivalents in Scotland, Wales and North Ireland.

If such a certificate is later revoked, it is treated as never having been issued.

If only some components of an item of plant qualify for first-year allowances under this heading, the amount qualifying is limited to the amount specified by Treasury Order.

The Treasury Order giving effect to these provisions (the Capital Allowances (Energy-saving Plant and Machinery) Order 2001, S1 2001 No. 2541) operates by giving statutory authority to the energy technology lists issued by the Department for the Environment, Food and Rural Affairs and HM Revenue and Customs, and managed by the Carbon Trust. Plant or machinery qualifies as energy-saving plant or machinery, therefore, if

- it falls within a technology class specified in the Energy Technology Criteria List;
- it meets the energy-saving criteria set out in that List;
- in case of classes (4) to (12) below, it is of a type specified in (and not removed from) or which has been accepted for inclusion in, the Energy Technology Product List.

Initially, the Energy Technology Criteria List specified the following technology classes:

1 combined heat and power;
2 lighting;
3 pipework insulation;
4 boilers;

5 motors and drives;
6 refrigeration;
7 thermal screens.

With effect for expenditure incurred on or after 5 August 2002, the following further classes were added:

8 heat pumps;
9 radiant and warm air heaters;
10 compressed air equipment;
11 solar thermal systems.

With effect for expenditure incurred on or after 5 August 2003, the following further class was added:

12 automatic monitoring and targeting equipment.

With effect for expenditure incurred on or after 26 August 2004, the following further classes were added:

13 air to energy recovery equipment;
14 compact heat exchangers;
15 heating, ventilation and air-conditioning zone controls.

The Energy Technology Criteria and Product Lists can be viewed on the ECA website at www.eca.gov.uk.

Where expenditure was incurred, or a contract entered into, on or after 1 April 2001 but before the date of the making of the Treasury Order, the plant or machinery concerned can nevertheless be energy-saving plant or machinery provided that, at the time the expenditure is incurred or the contract is entered into, it meets the conditions specified in the Order.

Additions to the types of qualifying technologies were set out in Statutory Instruments SI2002/1818, SI2004/2093 and SI2005/2424.

The Finance Act 2003 introduced a 100% first-year allowance for expenditure on environmentally beneficial plant and machinery incurred on or after 1 April 2003.

The plant acquired must be of a description specified by Treasury Order, and meet the environmental criteria set out by the Treasury Order. In essence, 'environmental beneficial' plant will have been designed to remedy or prevent damage to the physical environment or natural resources CAA 2001, s45 H(3). In some cases, allowances are due only if a 'certificate of environmental benefit' is in force, confirming that particular plant (or plant constructed to a particular design) meets the relevant environmental criteria.

The plant acquired must be new and not second-hand, and must not be a long-life asset (i.e., having an estimated useful economic life of greater than 25 years).

Although the allowances for environmentally beneficial plant and machinery are similar to those for energy-saving plant and machinery CAA 2001, s45A, the latter may be leased, whereas the former may not.

Again, allowances are available for environmentally friendly components of larger plant, even if the plant as a whole would not qualify. The amounts qualifying are those specified by the relevant Treasury order, and not a just apportionment of the whole cost – CAA 2001, s45J.

If a certificate of environmental benefit is revoked, it is regarded as never having been issued, and tax returns must be amended accordingly. A taxpayer who becomes aware that amendment is needed has three months in which to notify HM Revenue and Customs.

Additions to the types of qualifying technologies were set out in statutory instruments SI2004/2094 and SI2005/2423.

Industrial buildings

Industrial buildings are defined as 'buildings used for the qualifying trades stated in the Capital Allowances Act 2001. It is the use of the building that is important. Qualifying uses most commonly include buildings which are used for manufacturing and processing goods; mills, factories, etc., buildings used for the storage of materials to be used in the manufacture of other goods; and buildings used for the repair and maintenance of goods (but generally not forming part of a retail business).

Allowances for industrial buildings are presently 4% per annum calculable on a straight-line basis from the time the building was first used as an industrial building. The allowances are available in respect of the cost of construction. This excludes the cost of purchasing the land, but does include the cost of site preparation. For the purposes of capital allowances, industrial buildings have a maximum life of 25 years; after this period of use as an industrial building no allowances are available. For buildings which were brought into use prior 6 November 1962, the writing-down allowance is 2% per annum and the tax life is 50 years. Initial allowances have, in the past, been available for the first year of use of a qualifying building.

If a qualifying building is sold during its tax life, a balancing adjustment is made in respect of allowances claimed by the vendor. Depending on the sale price, this could result in a balancing allowance or a balancing charge. The purchaser is entitled to allowances for the remainder of the building's tax life. The calculation of the relief available is, however, based upon the original construction cost of the building. Example 2 illustrates the position.

Example 2

In December 1998, the construction costs of an industrial building were £1,000,000, excluding associated land costs of £500,000. Annual allowances totalling £320,000 have been claimed for eight years' use. The building was sold in 2006 for £1,100,000.

The position with regard to industrial building allowances is as follows:

	£
Proceeds of sale	1,100,000
Less land cost	500,000
Cost of building	600,000
Original building cost	1,000,000
Cost of owning building	400,000
Allowances claimed	320,000
Vendor's balancing allowance	80,000

The remaining tax life of the building (ignoring months for simplicity) is 17 years. The purchaser would get the following annual relief:

Annual relief = £600,000/17 years

If the building had sold for £1,400,000 the position would be as follows:

	£
Proceeds of sale	1,400,000
Less land cost	500,000
Cost of building	900,000
Original building cost	1,000,000
Cost of owning building	100,000
Allowances claimed	320,000
Vendor's balancing charge	220,000

The remaining tax life of the building (ignoring months for simplicity) is 17 years. The purchaser would get the following annual relief:

Annual relief = £900,000/17 years

Hotels

Capital allowances are available against the construction cost of hotels. To qualify for relief, a hotel must be a permanent building which is open for at least four months between April and October, offer at least ten letting bedrooms and provide normal hotel services (i.e., breakfast, room cleaning, etc.). The writing-down allowance of 4% per annum based upon initial construction cost is calculated on the same basis as that for industrial buildings. The situation in respect of balancing adjustments on the sale of a hotel building is also dealt with in a similar way to the adjustments for industrial buildings.

Enterprise zones

Enterprise zones are areas so designated by the Secretary of State. Expenditure on commercial and industrial buildings within enterprise zones, but excluding expenditure on dwelling houses, is eligible for 100% initial allowance (however, a lower initial allowance can be claimed). Expenditure either incurred or contracted for within 10 years of the creation of an enterprise zone qualifies for relief. Buildings which are sold within the first 25 years are subject to the balancing adjustments described above for industrial buildings.

Financial impact for a development

A single development can attract more than one type of capital allowance. It is therefore important to plan any claim for relief to ensure that the maximum benefits will accrue. For example, allowances for a hotel development could be claimed on the basis of 4% writing-down allowance on the whole construction cost. However, it may be more financially attractive to extract the qualifying expenditure on plant and machinery and ECA's from the cost of the hotel and claim these separately on a 25% reducing balance basis and 100% respectively (ECA's), thus receiving the relief sooner. The benefits of different tax-planning strategies will be different for individual clients, and so tax planning should take account of a client's overall business status and objectives.

The following table provides typical ranges of the levels of plant and machinery allowances often achieved for a range of different property types. These percentages should be applied to the capital expenditure incurred, however, please note that the entitlement to claim does not always exist and should be researched on a project by project basis.

Development type	*From (%)*	*To (%)*
Offices		
Low to medium rise, not air conditioned	20	30
Air conditioned	25	35
1st Class	30	45
Fitting out shell office	55	85
Industrial		
Basic Units	2	5
Office/Hi-Tech units	5	25
Retail		
Shop Shells	2	25
Department Stores	15	35
Shopping Centres	10	30
Fit-out Existing Retail units	55	85
Distribution Warehouses	5	15
Leisure		
3–5 Star Hotels	20	65
Leisure Centres	20	40
Pubs and Restaurants	20	50
Fit-out Existing Pub	55	90
Healthcare		
Hospitals	30	45
Nursing Homes	25	35
Health Centres/Surgeries	25	35
Specialist Buildings		
Computer Centres	80	95

*Contaminated Land Relief offers 150% tax relief on qualifying expenditure.

†ECA provides 100% tax relief on qualifying expenditure.

Location of Enterprise zones

The complete list of zones created, together with their expiry dates, is as follows:

Lower Swansea Valley (No. 1)	10 June 1991
Corby	21 June 1991
Dudley	9 July 1991
Langthwaite Grange (Wakefield)	9 July 1991
Clydebank	2 August 1991
Salford Docks	11 August 1991
Trafford Park	11 August 1991
City of Glasgow	17 August 1991
Gateshead	24 August 1991
Newcastle	24 August 1991
Speke	24 August 1991
Hartlepool	22 October 1991
Isle of Dogs	25 April 1992
Delyn	20 July 1993
Wellingborough	25 July 1993
Rotherham	25 September 1993
Scunthorpe (Normanby Ridge and Queensway)	15 August 1993
Dale Lane and Kinsley (Wakefield)	22 September 1993
Workington (Allerdale)	3 October 1993
Invergordon	6 October 1993
North West Kent (Nos. 1–5)	30 October 1993
Middlesborough (Britannia)	7 November 1993
North East Lancashire	6 December 1993
Tayside (Arbroath)	8 January 1994
Tayside (Dundee)	8 January 1994
Telford	12 January 1994
Glanford (Flixborough)	12 April 1994
Milford Haven Waterway (North Shore)	23 April 1994
Milford Haven Waterway (South Shore)	23 April 1994
Dudley (Round Oak)	2 October 1994
Lower Swansea Valley (No. 2)	5 March 1995
North West Kent (Nos. 6–7)	9 October 1996

(*continued*)

Inverclyde	2 March 1999
Sunderland – Hylton and Southwick	26 March 2000
Sunderland – Castleford and Doxford Park	26 March 2000
Lanarkshire (Hamilton)	31 January 2003
Larnarkshire (Monkland)	31 January 2003
Larnarkshire (Motherwell)	31 January 2003
Dearne Valley (Nos. 1–6)	2 November 2005
East Midlands (NE Derbyshire)	3 November 2005
East Midlands (Bassetlaw)	16 November 2005
East Midlands (Ashfield)	21 November 2005
East Durham (Nos. 1–6)	29 November 2005
Tyneside Riverside (No. 1)	19 February 2005
Tyneside Riverside (Nos. 2–7)	26 August 2006
Tyneside Riverside (Nos. 8–10)	21 October 2006

Rates of allowances

Buildings

Initial allowances	%
Industrial buildings, commercial buildings and structures in enterprise zones	100
Industrial buildings, etc., in other geographical areas	Nil
Buildings used for scientific research	100
Conversion of business premises into flats	100
Writing-down allowances	
Industrial buildings, commercial buildings and structures in enterprise zones (when initial allowance is disclaimed)	25
Industrial buildings	4
Commercial (non-industrial) buildings not in an enterprise zone	Nil
Qualifying hotels	4
Agricultural buildings	4

Machinery and Plant

First-year allowances (all enterprises)	
Expenditure incurred on or after 1 April 2001 on energy-saving plant or machinery	100

Expenditure incurred on or after 1 April 2003 on environmentally beneficial plant or machinery	100
First-year allowances (for small or medium-sized enterprises only)	
Expenditure incurred 2 July 1997 to 1 July 1998	50
Expenditure incurred 2 July 1997 to 1 July 1998 (long-life assets)	12
Expenditure incurred on or after 2 July 1998	40
First-year allowances (for small enterprises only)	
Expenditure incurred by small enterprises on information and communications technology, in four years ending 31 March 2004	100
Expenditure incurred in 12 months beginning 1 April 2004 (6 April 2004 for income tax)	50
Writing-down allowances	%
Long-life assets	6
Plant used for overseas leasing	10
Other plant (NB: writing-down allowance on expensive cars restricted to £3000 per annum)	25
Land remediation	
Expenditure incurred by a company on or after 11 May 2001	150
Research and Development	
Capital expenditure	100
Revenue expenditure incurred by a small or medium-size enterprise on or after 1 April 2000	150
Revenue expenditure incurred by a large company on or after 1 April 2002	125

Bibliography

1. WILSON, M. and BONE, S., *Tottel's Capital Allowances: Transactions & Planning 2005–06* (Tottel Publishing, 2005).
2. WALTON, K. and SMAILES, D., *Tolley's Capital Allowances 2005–06* (Reed Elsevier, 2005).

14

Indemnity and insurance

Introduction

The risks of damage and injury associated with construction projects are often high. Insurance provision is therefore an essential requirement in construction contracts to protect the interests of those involved. It is, without question, necessary for those engaged in advising clients and administering contracts to have a basic knowledge of the principles of indemnity and insurance, together with an understanding of the provisions and operation of the associated clauses in standard forms of building contract. This knowledge and understanding will help to ensure that

1 Clients are informed about the responsibilities for insuring risks under a contract, thus enabling them to comply with any obligations imposed upon them by the contract conditions and to arrange additional cover where necessary.
2 The most suitable insurance clauses (from the various alternative standard clauses) for individual projects are used and an adequate level of financial cover is stipulated within contract conditions.
3 The insurance policies put in place for a contract comply with the requirements for insurance cover in the contract conditions.
4 The correct procedures are adopted in the event of an insurance claim associated with loss or damage.

This chapter provides a basic introduction to indemnity and insurance associated with building contracts. In particular, it focuses on the provisions of the Standard Building Contract with

Quantities, 2005 Edition,[1] and references to contract clauses in this chapter relate to this contract.

Some general principles of insurance

There are two basic types of insurance

1 *Liability insurance*, providing financial cover for legal liabilities owed to others.
2 *Loss insurance*, providing cover for losses, which fall directly on the insured parties.

A contract of insurance is a legal contract, and therefore the general criteria relating to the existence of a contract pertain. To take out insurance cover, a person or organization must have an insurable interest (e.g., ownership of a house). The principle of utmost good faith requires the insured to disclose to the insurers all material information in respect of the risks being insured. The principle of indemnity means that insurance will put the insured back into the position they would have been in, had the risk not occurred (i.e., no loss and no gain). *Subrogation* is the insurers' right to pursue claims that the insured may have against third parties in respect of the insurers' indemnification.

'Standard' insurance policies do not exist. It is therefore essential to check that the insurance provided by a policy gives the protection required. This can be difficult, as many of the terms used in insurance policies can be open to different interpretations; the plethora of case law on this subject highlights the problem. It should also be borne in mind that the insurance market is not static; insurance policies and products are continually being updated to take account of new technologies and new risks – policies must therefore be carefully scrutinized.

Indemnity

Clause 6 places obligations upon contractors to indemnify Employers against the consequences of injury to persons and property.

Clause 6.1 requires contractors to indemnify Employers in the event of personal injury or death to any person arising out of, or in the course of, or caused by, the carrying out of the Works. A contractor's liability under clause 6.1 is for any expense, liability, loss, claim or proceedings against an Employer arising under any statute or at common law. A contractor's liability does not, however, extend to injury or death arising as a result of an act or neglect on the part of an Employer or of persons for whom an Employer is responsible (including persons directly employed as defined by clause 2.7).

Clause 6.2 requires contractors to indemnify Employers in respect of any expense, liability, loss, claim or proceedings arising from injury or damage to property, real or personal. Injury to property could include situations where, although there may not have been physical damage, enjoyment of a right, for instance, may have been affected – for example, an easement or a loss of light caused as a result of construction activity. This indemnity is limited to injury or damage arising out of, or in the course of, or by reason of, the carrying out of the Works. Indemnity is further limited to injury or damage to property due to negligence, breach of statutory duty, omission or default of 'the Contractor or any of the Contractor's Persons' (see clause 1.1, Definitions for an explanation of Contractor's Persons).

The contractor is not, however, responsible for injury or damage to property caused by the 'Employer or any person employed, engaged or authorised by him or by any local authority or statutory undertaker executing work solely in pursuance of its statutory rights or obligations'. Where a statutory body is acting as a domestic sub-contractor, then the contractor will be liable for injury or damage caused by them under clause 6.2. 'Property real or personal' excludes the Works, any work executed, and materials on site. Damage to the Works etc., is covered by the insurance provisions of clause 6.7; clause 6.2 deals with indemnifying the Employer against damage to third party property.

Where work is to existing structures (for example, repair, alteration or refurbishment projects) and Insurance Option C applies with regard to insurance of the Works, the indemnity required by clause 6.2 excludes injury and damage to the existing structures and their contents resulting from Specified Perils. Damage caused by Specified Perils is dealt with in clause Insurance Option C.1. However, indemnity is required for injury or damage caused other than by a Specified Peril. The definition of Specified Perils is given in clause 6.8 as 'fire, lightning, explosion, storm, flood, escape of water from any water tank, apparatus or pipes, earthquake, aircraft and other aerial devices or articles dropped therefrom, riot and civil commotion, but excluding Excepted Risks'.

Insurance against injury to persons and property

The requirements

Clause 6.4 provides a financial 'backbone' to the indemnities given to the Employer under clauses 6.1 and 6.2. It places a responsibility upon contractors to effect insurance cover against their liabilities under the contract. Insurance is described by Madge[2] as 'a risk-spreading mechanism, spreading the losses and liabilities of the few amongst the many'. The cost of insurance will, to a large extent, be related to the types of risks being taken; the more claims and awards there are across the industry, the higher will be the cost of insurance. This cost is passed on to the clients of the construction industry through the contracts awarded.

The insurances required by clause 6.4 are to cover personal injury and death to third parties and damage to property (other than the Works). Clause 6.4 is an insurance clause, as opposed to an indemnity clause; it places a contractual obligation on contractors to arrange insurance. It does not relieve contractors of their obligation to indemnify the Employer under clauses 6.1 and 6.2.

Clause 6.4.1.1 requires insurance for personal injury or death to a contractor's employees to be in accordance with the Employer's Liability (Compulsory Insurance) Act 1969. This insurance cover is a statutory requirement, and therefore the clause acts as a reminder to contractors of their statutory obligations. Other insurance required by the clause is to be sufficient to cover all other liabilities imposed upon a contractor by clauses 6.1 and 6.2. The financial cover provided must be not less than the sum stated in the Contract Particulars to the contract 'for any one occurrence or series of occurrences arising out of one event'. The insurance cover is required to indemnify the Employer as well as the contractor. It is not practicable for Employers to require unlimited indemnity; insurance companies do not provide this cover. The minimum amount of insurance cover required must therefore be established by the Employer. The amount will depend upon the type, size, location, etc., of the project. Practice Note 22[3] advises that the amount stated must be realistic and, if necessary, advice should be sought from the insurance experts in establishing the figure. However, a contractor must be aware that this amount is merely the minimum required by the Employer. The contractor will be liable for any claims that may exceed this amount and should therefore consider taking out cover in excess of the minimum where appropriate.

Documentary evidence

Clause 6.4.2 enables Employers to elicit documentary evidence from contractors to the effect that insurances required in respect of third party cover have been put in place and are being maintained. The procedures required by the contract are for documents to be sent to the architect (upon the request of the Employer) for inspection by the Employer. It is common, in practice, for Employers to request the architect, or design team, to check that these documents are in accordance with the requirements of the contract; indeed, where Employers are unfamiliar with construction contracts, the design team should advise them of their contractual right to request evidence in order to protect their interests.

Insurance is a complex field, and checking that the provisions of individual insurance policies comply with the requirements of building contracts can be fraught with dangers. Faced with such a request, it is important to take the correct course of action. This will depend upon whether an individual or firm has the necessary insurance expertise in each particular situation (some situations may be more straightforward than others) to verify the adequacy of cover, or if specialist advice is required. Bearing in mind the potentially large sums that may be involved in the event of an insurance claim, such decisions must be made carefully (see *Pozzolanic Lytag v Bryan Hobson Associates* in Chapter 1).

Length of cover

The contract is not explicit with regard to the length of time for which this insurance should be maintained. Practice Note 22[4] gives some guidance in that insurance taken out in compliance with clause 6.4 should not be terminated 'until at least the expiry of the Defects Liability Period or the date of issue of the Certificate of Completion of Making Good Defects whichever is the later'. This view is supported by Madge:[5] 'such insurances ought to remain in force at least to the end of the defects liability period'. Under the new terminology adopted by JCT 05, this now equates to the expiry of the Rectification Period or the date of issue of the Certificate of Making Good.

Default by a contractor in effecting insurance cover

If a contractor defaults in taking out or maintaining the insurance prescribed by the contract, there is a remedy for the Employer (see clause 6.4.3). Employers are empowered, in such circumstances, to take out insurance against any liability or expense, which could arise as a result of such a default. The cost of this insurance can either be deducted from money due to the contractor from the Employer or, alternatively, can be recovered by the Employer from the contractor as a debt. It should be noted that, in these circumstances, Employers would only insure their own liability – they are not required to take out cover to protect a contractor against liability claims (in fact, it would be difficult for an Employer to arrange such cover).

Excepted risks

Contractors are not required to indemnify Employers or to insure against injury, death or damage resulting from Excepted Risks. Excepted Risks are defined in clause 6.8 and include nuclear perils and the effects of sonic or supersonic aerial devices. These risks are generally excluded from the cover provided by insurance policies and hence their exclusion from the contractual requirements to insure. Liability for injury or damage resulting from these risks is generally covered by other mechanisms, such as nuclear legislation (where liability is that of licensed nuclear operators) and the UK government's undertaking to compensate for damage arising out of sonic bangs, which would have covered any damage caused by Concorde during the time it was in service.

Damage to property not covered by a contractor's indemnity

Indemnity is provided by a contractor under clause 6.2 only with regard to injury or damage to property due to negligence, breach of statutory duty, omission or default on the part of the contractor (or those for whom that contractor is responsible). Clause 6.4 requires contractors to arrange insurance to support this indemnity.

The situation could arise, however, whereby property is damaged as a consequence of construction work but not as a result of negligence, etc., on the part of the contractor. An Employer could be held liable for such damage. Typical examples of situations resulting in such damage include subsidence, heave and the lowering of ground water. JCT 05 provides an optional provision in clause 6.5 by which the liability of the Employer in such circumstances can be protected by insurance cover. This clause does not require contractors to indemnify Employers against risks, but obligates them to arrange a joint insurance on behalf of an Employer to cover the Employer's liability.

If it is anticipated that this insurance will be required, a statement should be included in the Contract Particulars to this effect (i.e., that insurance may be required) together with the amount of indemnity required. Decisions regarding whether this insurance will be necessary in a particular circumstance and, if so, the amount of insurance cover required will depend upon the particular circumstances of the Employer, the project and the site. If insurance is required, the architect must instruct the contractor to take out a Joint Names Policy for the amount of indemnity stated in the Contract Particulars. The policy is required to indemnify the Employer (although it is in joint names) for any expense, liability, loss, claim or proceedings against the Employer resulting from injury or damage to property (excluding the Works and materials on site) 'caused by collapse, subsidence, heave, vibration, weakening or removal of support or lowering of ground water' as a result of the construction work.

The following items are specifically excluded from the above insurance – injury or damage, which is either

(i) a contractor's liability under clause 6.2; or
(ii) caused by design error; or
(iii) an inevitable consequence of the construction work; or

(iv) is the responsibility of the Employer to insure under clause Insurance Option C.1 (i.e., damage caused by Specified Perils to existing structures); or
(v) to the Works and site materials, except where a practical or sectional completion certificate has been issued; or
(vi) arising from war, invasion, act of enemy hostility, civil war; or
(vii) arising from Excepted Risks; or
(viii) arising directly or indirectly from pollution or contamination, with some exceptions; or
(ix) the result of any costs being incurred by the Employer by way of damages for breach of contract, except where such costs would still have arisen in the absence of any contract.

The cost of this insurance should be added to the Contract Sum. The insurers used by a contractor to provide the cover required by this clause must be approved by the Employer. Policies and premium receipts are to be sent by the contractor to the architect for depositing with the Employer.

The contract makes no mention of what action should be taken if a contractor defaults in taking out or maintaining the insurance cover as instructed. In such circumstances an Employer should be advised to take out such cover as has not been effected and recover any outstanding premium payments that may have been passed to the contractor.

Insurance of the works

Clause 6.7 deals with the provisions for the insurance of the Works and provides the Employer with a number of choices, i.e., Insurance Options A, B and C which are set out in Schedule 3 of the contract. The general obligation is that All Risks insurance must be provided, giving cover for physical loss or damage to work executed and site materials (but with some exclusions).

Insurance Options A, B and C are the three alternative clauses, which provide the mechanisms for insurance of the Works. Only one of these clauses will be applicable for any one contract. Insurance Option A covers the insurance requirements for new buildings where the contractor is made responsible for arranging the insurance; Option B is for new buildings for which the Employer is responsible for arranging the insurance; and Option C should be used where work is in, on or to, existing structures – it requires insurance of the existing structure and contents and insurance of the Works to be arranged by the Employer. The applicable clause for a contract must be stated in the Contract particulars.

The policy of insurance of the Works must be in joint names to ensure that protection is provided to the parties against subrogation. The all risks policy is to make provision for sub-contractors but only to the extent of the Specified Perils element of the policy. Therefore, sub-contractors are protected from injury and claims arising from specified perils but are not covered to the full extent of the all risks policy. This protection is to be provided either by naming these sub-contractors or by a waiver in the insurance policy of the right of subrogation, which an insurer may otherwise have against a sub-contractor. Sub-contractors will have to arrange their own insurance for loss or damage caused other than by Specified Perils.

All Risks insurance must be maintained up to the date of issue of the certificate of Practical Completion or the date of termination of the employment of the contractor, whichever is the earlier. It is essential at this point to ensure that clients are aware of the date on which insurance of the Works ceases to enable them to arrange insurance cover for the buildings to take effect on the cessation of the Works insurance.

The insurance requirements of clauses 6.7, Insurance Options A, B and C do not replace the obligations placed upon contractors by clause 2.1 to carry out and complete the Works. However, while the risk may lie with them, contractors maintain the right to take legal action against third parties who are liable for loss or damage to the Works. It is in both the contractor's and the Employer's interests that insurance is taken out in joint names. This ensures that a contractor's financial exposure in the event of loss or damage is covered, which provides a safeguard for both the parties. Employers have an insurable interest in the Works by virtue of interim payment procedures. Joint insurance relieves Employers of any liability they may have to an insurance company for any responsibility they may have for loss or damage caused to the Works.

Insurance Option A – All Risks insurance of the Works by the contractor – New buildings

Insurance Option A should be used for projects involving the construction of new buildings and in situations where the Employer does not have (or does not wish to arrange) insurance cover for the work in progress. Traditionally, it was the contractor's obligation to arrange Works insurance, but there has been a move away from this situation by Employers who gain benefit from arranging insurance themselves (for example, Employers who are frequent developers might be able to purchase insurance at a competitive price). In such circumstances, Insurance Option B should be used (but normally should only be contemplated at the express wish of an Employer).

Insurance Option A requires contractors to take out a Joint Names Policy for All Risks insurance to cover the full reinstatement value of the Works. The Contract Particulars are to state what percentage is required to cover professional fees associated with the reinstatement of the Works, if this is left blank then a figure of 15% is to apply and the insurance cover should include this percentage. The policy must be maintained until the earlier of either

(i) the date of issue of the certificate of Practical Completion; or
(ii) the date of termination of the employment of the contractor.

There are two alternative ways that cover may be provided by the contractor, either:

(i) a bespoke policy for the project, placed with insurers approved by the Employer (Insurance Option A.1). The contractor is required to send the insurance policy and premium receipts to the architect to be deposited with the Employer; or
(ii) if the contractor has an annual insurance policy which provides All Risks insurance then it may be used, provided it complies with the requirements of the contract concerning the provision of All Risks insurance (Insurance Option A.3). The contractor must make the policy documentation available for inspection by the Employer if requested. The renewal date of annual policies must be stated in the Contract Particulars, thus enabling the continuity of cover to be checked by the Employer at the appropriate times.

If a contractor fails to take out or maintain insurance in accordance with the requirements of the contract, the Employer is empowered to take out Joint Names cover in respect of the default. The cost of this insurance can either be deducted by the Employer from sums owing or recovered from the contractor as a debt.

If loss or damage occurs as a result of an insured risk, the following action should be taken:

1 The contractor must give written notice to the architect and the Employer, providing details of the extent, location and nature of the loss or damage.
2 No account should be taken of the loss or damage when calculating payments due to the contractor, e.g., where a fire may have substantially damaged work carried out by the contractor the full value of the original work is to be included in any subsequent interim valuation.
3 Once the insurers have finished their inspections of the loss or damage, the contractor should restore, replace and repair as necessary and proceed with the Works. It is advisable to obtain express notice from the insurers that they have finished inspections before starting the restoration of damage.
4 The contractor must authorize the insurers to pay all sums due under the insurance claim to the Employer.
5 The contractor is paid for the reinstatement work by the Employer in instalments as certified by the architect. These certificates should be issued at the normal period of interim certification. The contractor is entitled to receive the full sums received from the insurers (excluding sums for professional fees), but is not entitled to receive any sums beyond those received under the insurance for reinstatement. Therefore, any shortfall must be made good by the contractor.

Insurance Option B – All Risks insurance of the Works by the Employer– New buildings

If Insurance Option B is used, All Risks insurance is to be arranged by the Employer in Joint Names. The clause requires the Employer to provide documentary evidence of the insurance when reasonably required to do so by the contractor. If the Employer defaults in taking out or maintaining insurance, the contractor can take out Joint Names cover in respect of the default and the cost is added to the Contract Sum.

The procedures in the event of loss or damage under Insurance Option B are similar to those prescribed in Option A, except with regard to payment to the contractor. Under Insurance Option B, reinstatement works are treated as though they were a variation required by the architect under clause 3.14.1. Payment is not therefore governed by the limit of the monies received from an insurance claim. If the insurance monies do not adequately cover the cost of repairs, etc., then the Employer is responsible for any shortfall.

Insurance Option C – Insurance of existing structures and insurance of the Works in, or extensions to, existing structures

This clause is to be used when construction involves work to an existing building (for example, refurbishment, repair, fitting-out, extensions, etc.). There are two separate issues to deal with here:

1 Insurance of the existing structure and its contents.
2 Insurance of the Works.

The Employer is required to arrange a Joint Names Policy in respect of the existing structure and its contents (Insurance Option C.1). The insurance must provide cover for the full cost of reinstatement, repair and replacement of loss or damage caused by the Specified Perils. This insurance must be maintained until either the date of issue of the certificate of Practical Completion or the date of termination of the employment of the contractor, whichever is the earlier. In the event of a claim, the contractor is required to instruct insurers to pay all sums to the Employer.

The requirements for insurance of the Works, and the procedures and provisions with regard to the occurrence of loss or damage, are similar to those under Insurance Option B (as described above). However, there is an additional provision in Insurance Option C.4.4 whereby a right is given to either party (if just and equitable) to terminate the employment of the contractor within 28 days of the loss or damage.

The remedies for the contractor, in the event of an Employer failing to take out and maintain insurance as required, are the same as those in Insurance Option B.

General note on Insurance Options A, B and C

There may be circumstances where it is impossible to arrange the complete insurance cover prescribed by the contract for a project (for example, in areas where there is a high risk of flooding, there may be exclusions relating to flood damage). In such circumstances, the footnotes in the contract suggest that arrangements should be agreed between the parties and the Contract Conditions amended accordingly.

Professional indemnity insurance

Professional indemnity insurance, as its name implies, is designed to provide cover for those who provide a professional service. Such insurance is normally associated with professionals, such as architects, chartered surveyors, engineers, etc. Some of the professional institutions, e.g., the RICS, insist that their members carry professional indemnity insurance if they are offering a professional service under their chartered status. On the face of it there would appear to be little need for any contracting organization to be concerned with this area of insurance, but this may be a false assumption, for many contractors are expanding their design expertise to provide specialist advice to the client and his advisers. In many instances they will be employed on a design and build basis, whereby they provide a detailed design scheme for the client's project and install it.

By carrying out these tasks, the contractor is taking on the traditional role of the professional designers and accepting their associated liabilities, or possibly even greater liabilities. It is a quirk of the English legal system that, where a 'professional' prepares a design, it is normally implied that the design will have been prepared with 'reasonable skill and care' but, when a contractor carries out design and build work, it will normally be implied that the work and design shall be 'fit for its purpose'; a far more onerous liability than that accepted by a professional. It follows that, where a contracting organization carries out design work or gives specialist advice as a normal part of their works, it is essential that they have professional indemnity insurance and this fact should be checked by the Employer or his advisers.

A professional indemnity policy will indemnify the policyholder against claims made for a breach of professional duty, i.e., failing to exercise reasonable skill and care in designing the works or in the choice of materials and components. The policy will not, in the case of a contractor, provide cover for bad workmanship or faulty materials. It is also unusual for the policy to cover a 'fitness for purpose' liability, therefore it is important for any design and build contractor to carefully examine the contract documentation to identify the level of liability imposed, i.e., fitness for purpose or reasonable skill and care. Under the JCT form a contractor is expected to accept the liability of an independent professional designer, i.e., to exercise reasonable skill and care.

If a contractor were to sub-let the design work to an outside professional organization, he would still be required to maintain professional indemnity insurance as he is contractually liable to the employer for the adequacy of that design. However, although the initial responsibility for the design work will rest with the contractor he may be able to pass the responsibility down the line to the professional design team, in which case it is essential that the contractor checks to ensure that the designer has adequate insurance and maintains it for as long as necessary.

Professional indemnity insurance is annually renewable and will have a monetary limit imposed upon the level of indemnity it provides. The limit may be for the aggregate of claims made during the year or for each individual claim. One of the problems with this type of insurance is that it operates on the principle of claims made. This means that it is the insurer who is currently providing cover for a professional or contractor on the date when a claim is made who will be responsible for the costs, and not the insurer who provided the cover at the time the negligent work was carried out. Therefore, although it is important for the client to know that a designer has professional indemnity insurance at the time the work is executed, it is perhaps even more important for him to know that the insurance will be maintained for perhaps a further 12 or 15 years after completion. It is difficult for the client to enforce this requirement, although it can prove to be a very expensive mistake for any professional to ignore this requirement.

The danger of not maintaining appropriate insurance cover was highlighted in the case *of Merrett v Babb (2001) 80 ConLR 43*. A firm of surveyors and valuers employed a surveyor (Babb) as branch manager. Babb was an employee of the firm for approximately 11 months and it was during this time that he negligently prepared a mortgage and valuation report for a building society. A year or two after the purchase of the property the owner (Merrett) discovered that the property was suffering from settlement cracks and looked to recover compensation from the firm of surveyors and valuers who prepared the mortgage valuation report. This was an action in the tort of negligence because Merrett had no contract with the firm of surveyors. Because the owner of the firm of surveyors had been declared bankrupt, Merrett subsequently took action against the employee (Babb) who actually prepared the report and signed it off in his professional capacity as a chartered surveyor. The courts found Babb liable for making a negligent misstatement. An employee or ex-employee would normally expect the costs of being involved in such an action to be covered by his firm's professional indemnity insurance. Unfortunately for Babb the 'trustee in bankruptcy' of his old firm had cancelled the professional indemnity insurance and made no allowance for any 'run off' cover, which left Babb personally liable for all the costs.

Therefore, if surveyors, architects, contractors, etc., accept a design or other professional responsibility, they will need professional indemnity insurance. They must also be prepared to continue paying for this cover for a considerable time after completing the work until their legal liability has ceased. It is difficult for an organization to budget for such costs. The cost of such cover will fluctuate from year to year depending on the insurance market and the level of risk the insurers are prepared to accept.

JCT – contractor's design portion

Under the JCT 05 Form, it is possible a contractor may be required to take out and maintain a professional indemnity insurance policy. This procedure is only intended to be used where the contractor is responsible for an element of design within a project. Therefore, where the Employer or architect has decided to pass on some of the design responsibility to a contractor through the use of the Contractor Design Portion, it is the intention of the JCT that the contractor's design liability should be covered through insurance. Where a contractor is required to provide such insurance it is important to correctly complete the relevant details in the Contract Particulars. For example, within the Contract Particulars it is necessary to state the financial value of the insurance cover; if this section were to be left blank the default situation is that the insurance requirements of clause 6.11 will not apply.

The insurance requirements identified in clause 6.11 are if the contractor does not already have a professional indemnity policy in place he is required to take out the necessary insurance immediately after entering into the contract with the employer; the limit and amount of the cover is to comply with the requirements set out in the Contract Particulars; the insurance is to be maintained for the period set out in the Contract Particulars; and the contractor is to provide evidence that the policy exists and is being maintained.

Under clause 6.11 the contractor is required to maintain the professional indemnity insurance for either six or twelve years after the date of Practical Completion. The cost of professional indemnity insurance tends to fluctuate depending upon market conditions and the level of risk currently perceived by the insurance market. If the cost of maintaining the insurance becomes commercially unviable the contractor is to inform Employer, and both parties are then required to discuss alternative means of protecting their liability.

Joint fire code

In the early 1990s, the Insurance Market was hit with the high cost of fire losses on construction sites. For example, one insurance company[6] quoted the instance of just two sites, Minster Court and The Broadgate Centre, which together suffered somewhere in the region of £150m fire damage. It was in the light of such losses that the first Code of Practice for Fire Loss Prevention was introduced in May 1992. The fifth edition of 'The Joint Code of Practice on the Protection from Fire of Construction Sites and Buildings Undergoing Renovation' was published in January 2000.

If the joint fire code is to apply to a project, then this fact is confirmed by the appropriate entry in the Contract Particulars. The fire code that will apply to a project will be the one which is current at Base Date although, if the code is revised or updated during the progress of the project then the revised code will subsequently apply (Definitions, clause 1.1). It has been advised that the fire code will normally apply to all projects where the original value is

£2.5m or above although, depending upon circumstances, insurers may require the code to apply to projects of a lower value.[7] Where the fire code does apply, it is necessary to identify in the Contract Particulars whether or not the insurer has specified that the works is a 'Large Project'. According to the fire code, projects whose original value is £20m or above will be classified as large. Again, depending upon circumstances, it is possible that insurers may set a lower limit than this for a project. The implications of a 'large project' for the contractor or employer are the additional requirements of providing fire marshals and liaising with the emergency services about site plans, escape routes, etc. Obviously, where the contractor takes out the works insurance under Insurance Option A, the employer will have to rely upon the information supplied by the contractor when it comes to filling in this part of the appendix. The JCT guidance notes draw specific attention to a statement from the fire code warning that non-compliance with the code could lead to the All Risks insurance being withdrawn.[8]

Summary

The insurance requirements of a construction project can be very complex and difficult to understand. Failure to pay sufficient attention to this aspect of the works could lead to severe financial difficulties for all concerned. If a surveyor or contractor has any concern about the adequacy of any of the project insurances, then they should seek expert advice and clarification.

References

1. JCT, *Standard Building Contract with Quantities, 2005 Edition* (London: Sweet & Maxwell, 2005).
2. MADGE, P., *A Guide to the Indemnity and Insurance Aspects of Building Contracts* (London: RIBA Publications, 1985), p. 41.
3. JCT, *Practice Note 22 and Guide to the Amendments to the Insurance and Related Liability Provisions 1986* (London: RIBA Publications, 1986).
4. ibid., p. 34.
5. MADGE, P., *A Guide to the Indemnity and Insurance Aspects of Building Contracts* (London: RIBA Publications, 1985), p. 43.
6. Olympia-Axa Guide.

7. JCT, *Amendment 3 and Guidance Notes* (London: RIBA Publications, January 2001), Items 7–9, p. 10.
8. ibid., note, p. 9.

Bibliography

1. JCT, *Standard Form of Building Contract, Private With Quantities, 1980 Edition* (London: RIBA Publications, 1994).
2. MADGE, P., *A Guide to the Indemnity and Insurance Aspects of Building Contracts* (London: RIBA Publications, 1985).
3. JCT, *Practice Note 22 and Guide to the Amendments to the Insurance and Related Liability Provisions 1986* (London: RIBA Publications, 1986).
4. EAGLESTONE, F., *Insurance under the JCT Forms* (London: Collins Professional and Technical Books, 1985).
5. EAGLESTONE, F. N., *Insurance for the Construction Industry* (London: George Godwin Ltd, 1979).
6. RICS, *Introductory Guidance to Insurance under JCT Contracts* (London: RICS Books, 1991).

15

Health and safety: The Construction (Design and Management) Regulations 1994

Introduction

The Construction (Design and Management) Regulations 1994[1] (the Regulations) came into being by virtue of EU Directive 92/57/EEC on the implementation of minimum safety and health requirements at temporary or mobile construction sites. The Regulations are aimed at improving the health and safety management of construction projects. They consider all involved in the construction process (clients, designers, contractors and sub-contractors) and they make provision for safety to be considered for the whole life of construction projects from inception through design and construction, maintenance, adaptation and finally, demolition.

Surveyors should have a working knowledge of the Regulations to ensure that they comply with the duties imposed upon them by the Regulations. The term 'designer' in the Regulations refers to any person or organization that prepares a design; the term 'design' includes drawings, design details, specifications and bills of quantities. Much of the documentation produced by surveyors will therefore be classified as design. Surveyors may also be involved in the management of health and safety for a construction project and could, in some situations, be appointed to act as client and/or planning supervisor (a special duty holder created by the Regulations).

This chapter aims to provide a general understanding of the nature of the Regulations, to identify the duties placed on the various parties to the construction process by the Regulations, to explain when the Regulations apply to a construction project and to consider the relationship of the Regulations to other health and safety legislation.

The development of health and safety legislation

Health and safety law began life in the UK during the Industrial Revolution, when it became clear that some legal protection was required for employees to ensure their health and safety in the workplace. Early legislation, some of which is still applicable, was generally concerned with the provision of physical measures and prescriptive rules. There are many general and specific regulations that are applicable to construction, but it is beyond the scope of this chapter to identify and explain all the legislation.

A big step forward was made with the introduction of the Health and Safety at Work, etc., Act 1974. This Act created the Health and Safety Commission, which has the power to propose health and safety regulations and to approve codes of practice, and the Health and Safety Executive, which is responsible for enforcing health and safety legislation. Unlike previous health and safety legislation, the Act is concerned with improving the attitude of employers (and employees) to health and safety by providing a broad legislative framework incorporating general duties and instilling an organized approach to safety in the workplace. The Act is *framework legislation*; that is, it is the vehicle through which new regulations, including those resulting from EU Directives, are introduced.

Breach of health and safety legislation is a criminal offence; conviction can result in a fine and/or imprisonment.

'Reasonably practicable'

Many of the duties and requirements imposed by the Regulations are governed by what is reasonably practicable. 'Reasonably practicable' was explained by Asquith, L.F. in *Edwards v National Coal*

Board [*1949*, 1 KB 704] with regard to the duty of a mine owner to support the roof of a mine:

> 'Reasonably practicable' is a narrower term than 'physically possible', and implies that a computation must be made in which quantum of risk is placed in one scale and the sacrifice involved in the measures necessary for averting the risk (whether in money, time or trouble) is placed in the other, and that, if it be shown that there is a gross disproportion between them – the risk being insignificant in relation to the sacrifice – the defendants discharge the onus upon them. Moreover, this computation falls to be made by the owner at a point of time anterior to the accident.

It is therefore the subject of a cost-benefit exercise of assessing the cost of removing or controlling a hazard against the probability or severity of harm resulting from the hazard.

The construction industry's health and safety record

The construction industry in the UK is dangerous in comparison with other industries. Legislation has developed over the years in response to the analysis of data collected on the causes of accidents on construction sites. However, it is difficult for legislation to keep pace with the continuous and rapid developments in construction and materials technology.

The incidence of fatal accidents and accidents causing serious injury is consistently high. A report of the Health and Safety Executive in 1988 (Blackspot[2]) analysed the circumstances of 739 deaths in the construction industry between 1981 and 1985. The report considered that most of the deaths could have been prevented, and stated that 'better management of sites through detailed pre-site planning with all who are to be involved in the job is needed to improve the general level of safety. This requires discussion with architects, engineers and other professional advisers, as well as main and sub-contractors, safety representatives and safety professionals. Co-ordination of the work, with particular attention to high-risk activities, can reduce the overall risks'. The Construction (Design and Management) Regulations address these points; they consider the whole process of construction from initial design decisions through to maintenance and final disposal of buildings, place duties on all involved in the construction process, require the management and co-ordination of health and safety and require hazards to be identified and risks assessed.

The Regulations

The Regulations aim to ensure that health and safety is taken into account at all stages of the construction process. Construction involves many diverse activities. The employment and contractual arrangements across the industry and, indeed, for any one project, are complex and varied. The Regulations therefore focus on the management and co-ordination of activities to ensure a rational and collective approach to health and safety. They impose duties on clients, contractors and designers; they also create two health and safety management roles in the construction process; namely, the planning supervisor and the principal contractor. Joyce[3] notes, 'The planning supervisor will undoubtedly become an important source of advice to clients with regard to the Regulations'.

The Regulations have come into being in response to EU Directive 92/57/EEC on the implementation of minimum safety and health requirements at temporary or mobile construction sites which was adopted under Article 118A of the Treaty of Rome. The Regulations have been made under the Health and Safety at Work, etc., Act 1974. The Health and Safety Commission has given approval to the Approved Code of Practice HSG 224 (ACOP), which gives advice on how to comply with the Regulations. The legal status of the ACOP is such that if it is proved in a prosecution for breach of health and safety law that the provisions of the ACOP have not been followed, then a person or organization will be found at fault (unless it can be shown that the law has been complied with in some other way).

Application of the Regulations

The Health and Safety Executive is the enforcing authority for the Regulations. The Regulations are broadly applicable to all construction projects (with some exceptions), but they always apply to work involving demolition. If the construction phase of a project is expected either to be longer than 30 days or to involve more than 500 person/days then the Health and Safety Executive requires written notification of the project. The requirements on designers are applicable to all projects, irrespective of size or duration. Where a client believes that a project is not notifiable and will involve fewer than five persons at work carrying out construction at any one time, then the majority of the Regulations do not apply. In situations where the Local Authority is the enforcing agency for the construction work, the Regulations do not apply. When there is only one designer or one contractor associated with a construction project, the Regulations regarding co-ordination of design and construction respectively do not apply. If work is carried out for a domestic client, the majority of the Regulations do not apply; however, the Regulations place responsibilities on clients who commission work in connection with a trade, business or other undertaking. They also apply to speculative residential developers as if they were clients.

The health and safety plan

The health and safety plan is the medium through which health and safety associated with the construction of a project is managed and communicated. The health and safety plan will initially be prepared during the design phase of a project (often referred to as the pre-tender plan) and the planning supervisor has a duty to ensure that the plan is prepared. The plan is likely to be a collaborative effort, with information being provided by the client, the designers and the planning supervisor. The plan should be included with general tender information to contractors to enable them to include in their tender for dealing with the risks, issues and requirements of the plan.

The health and safety plan for a project is not a static document; it should be continuously reviewed, updated and amended as necessary. The principal contractor has a duty to develop the health and safety plan throughout the project (often referred to as the construction phase plan) and to support this there is a requirement (Regulation 19) for contractors to co-operate with the principal contractor, to comply with the rules of the health and safety plan and to provide information which might justify a review of the plan.

Information to be included at the preparation stage of the health and safety plan is detailed in Regulation 15(3). Possible information that could be included in a pre-tender stage health and safety plan is listed in Appendix 3 of the Approved Code of Practice HSG 224. This list includes details of the nature of the project; the existing environment; existing drawings; the design (including details of inherent hazards, work sequences, the principles of structural design and identification of specific problems for which the contractor will have to develop and propose risk management solutions); construction materials (associated hazards and precautions); site set-up (e.g., access, loading areas, traffic routes, etc.); site rules and requirements for continuous liaison.

The health and safety plan should be developed by the principal contractor before the construction starts and throughout the construction phase of the project. There is no requirement for the plan to continue after construction or for the principal contractor to hand over the plan on completion. The principal contractor is required by the Regulations to develop the health and safety plan. He should incorporate within the plan: the general framework for the management of health and safety for the project; risk assessments prepared by contractors in compliance with the Management of Health and Safety at Work Regulations (1999); rules for the management of construction required for reasons of health and safety; common arrangements (e.g., emergency procedures); details concerning co-operation, compliance with rules in the health and safety plan, authorized persons, the issuing of instructions and collection of information concerning health and safety; details explaining the requirements for providing information and training; and arrangements for allowing and co-ordinating the views of persons at work in respect of health and safety pertaining to the project.

The importance of the health and safety plan is clear: it is the prime tool for planning safety management for a project and all parties are involved in its development. The plan should provide a co-ordinated approach to the management of health and safety on site and it should incorporate methods for dealing with identified risks. It is likely that, when checking for compliance with the Regulations, the health and safety enforcement agencies will start by examining the health and safety plan, which could prove to be vital evidence in any decision to prosecute or, indeed, in a defence.

The health and safety file

The purpose of the health and safety file is to provide a comprehensive record of information pertaining to a building or structure to enable future persons involved in the maintenance, alteration and demolition of the building to design, plan and execute this work with due regard for health and safety. The health and safety file should be kept by the client and must be updated to take account of any changes that are made to the building or structure.

Typical information to be included in the health and safety file is listed in Appendix 4 of the Approved Code of Practice HSG224 and includes: record or 'as built' drawings and design criteria; general details of construction method and materials; details of equipment and maintenance facilities; specialist contractors' and suppliers' manuals giving details of operating and maintenance procedures; and details regarding the nature and location of utilities and services.

The client

Clients are required to determine if the Regulations apply to a project and, where appropriate, to appoint a planning supervisor and a principal contractor for all construction projects. The appointments are to be made as soon as practicable, and they should be reviewed, terminated and changed as necessary. Appointments must be in place throughout all projects up to completion of the construction phase. The Regulations permit the same person (or organization) to be appointed planning supervisor and principal contractor for a project (the principal contractor must be a contractor undertaking or managing construction work on the project) and they permit clients to appoint themselves as planning supervisor and/or principal contractor, providing they satisfy the requirements regarding competence and adequacy of resources.

Clients must give the planning supervisor information about the site which is relevant to the function of planning supervisor (Regulation 11). Such information is likely to include details of the existing buildings, land and associated plant. The health and safety file for a structure should be kept available by the client for inspection by any person who might need the information to comply with statutory requirements (Regulation 12). When disposing of an interest in a structure, a client must pass on the health and safety file to the relevant person or organization.

The Regulations impose a duty on clients to ensure that construction does not start until a health and safety plan for the project has been prepared (Regulation 10). It is anticipated that many clients may need to seek professional advice in making such a judgement, and it is imperative that sufficient time is allocated to principal contractors, prior to construction starting, to develop the plan. The ACOP recognizes that the use of some procurement methods will mean that there is often an overlap between the design, planning and construction phases of a project. The health and safety plan therefore needs only to be developed before construction may start, so far as the general framework for health and safety management of the project is concerned, together with those work packages which can reasonably be developed before the construction phase begins.

The duties imposed upon clients by the Regulations are onerous. They are permitted to appoint an agent to act as client in respect of the Regulations and therefore to undertake the duties imposed upon clients by the Regulations. This would be advisable action for clients with little or no experience of the construction process. Additionally, where there is more than one client for a project, one of the clients (or an agent) can elect to take the responsibility of the client under the Regulations.

The planning supervisor

Primarily, the planning supervisor has a co-ordination and advisory role. Duties imposed on the planning supervisor include: ensuring that designers comply with the Regulations and co-operate with other designers; being in a position to give advice to clients and contractors with regard to competency and adequacy of resources associated with engaging consultants, contractors and sub-contractors: ensuring that the health and safety file is prepared; reviewing, amending and adding to information in the health and safety file; and ensuring that the health and safety file is delivered to the client on completion of the construction work. The planning supervisor can be a company, a partnership or an individual.

The planning supervisor is required to give written notice of the project to the local office of the Health and Safety Executive (Regulation 7). The notice is to be given as soon after the appointment of the planning supervisor as is reasonably practicable. Information to be included in the notice includes: the address of the site; the names and addresses of the client, planning supervisor, principal contractor and any contractors chosen; the type of project; a declaration of the appointment of the planning supervisor, signed by or on behalf of the planning supervisor; a similar declaration for the principal contractor; the planned date for the start of construction; the anticipated construction duration; the estimated numbers of people at work on the site and the planned number of contractors on the site. The Executive has produced Form 10(rev), which can be used for giving notification (use of this form is not mandatory). Any information which is not available at the time the notice is given should be forwarded to the Health and Safety Executive as soon as it becomes available. Where the project is for a domestic client, the contractor is required to give written notice to the Executive before construction work starts.

Designers

The definition of 'designer' under the Regulations is wide, and includes activities that would perhaps traditionally not be thought of as design. Design includes drawings, design details, specifications and bills of quantities. The term 'designer' refers to any person who prepares (or arranges for persons under his or her control to prepare) a design. Thus surveyors, when performing some of their services, may be classified as designers within the meaning of the Regulations.

Designers must ensure that clients are aware of their (the clients') duties under the Regulations (Regulation 13(1)). They must include in their design adequate information about any aspect of the project that might affect the health and safety of those constructing and maintaining it. Designers are required to co-operate with the planning supervisor and other designers insofar as such co-operation will enable each of them to comply with statutory provisions.

Design work should be undertaken with a view to avoiding foreseeable risks to the health and safety of those carrying out associated construction, repairs and maintenance (and any person who might be affected by the work of such persons), combating risks to their health and safety and prioritizing health and safety measures which will protect all persons. Management of risks to health and safety is a key duty, and the onus is on designers to undertake risk assessment and to make design decisions which include consideration of the health and safety of those who will be constructing, maintaining, altering and demolishing the structures designed. It is not the intention of the Regulations that all risks should be

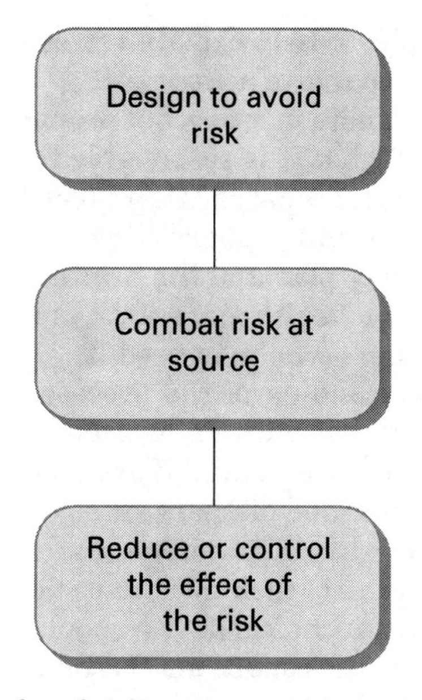

Figure 15.1 Hierarchy of risk management

avoided; decisions should be made by designers within the context of reasonable practicability. Reasonable practicability should be considered in terms of the costs of taking a course of action (measured in terms of finance, fitness for purpose, aesthetics, buildability and environmental impact) and the benefits that flow from that action.

The hierarchy of dealing with risks gives precedence to designing to avoid risks. If this is not possible, the causes of risk should be tackled at source and, failing this, the effects of risks should be reduced and controlled by protecting those persons whose health and safety might be affected by them (see Figure 15.1). The spirit of the Regulations in this respect is for designers to include in their decision-making process the effect that their design has on those who will be constructing and maintaining it. It has been recognized that decisions made during the design process can have an impact on health and safety. It is important to appreciate that this is about making judgements and exercising management. As with design, risk assessment and management should be a continuous process; health and safety issues should fall among all the other variables within the design decision process and should not be considered in isolation. The Regulations do not require designers to keep written records of health and safety considerations made during the design process, but there may be benefits associated with keeping records – for example, to help explain retrospectively the reasons behind a decision or course of action.

The duties on designers in respect of preparing designs are limited to the extent of what it is reasonable for a designer to do at the time the design is prepared. Designers have a duty to pass information to the planning supervisor so that it can be included in the health and safety plan and the health and safety file, and to provide information on health and safety aspects and other issues of their designs to those who may need it.

A practical understanding of the processes and techniques of risk assessment, together with an awareness of the construction implications of designs, is essential if designers are to make sound professional judgements and discharge their duties under the Regulations effectively. As with all questions of professional competence, designers must be aware of their own limitations and be able to recognize when specialist technical advice should be sought.

In situations where the conditions of a building contract require a contractor or sub-contractor to undertake some of the design work for a project, the Regulations, insofar as they apply to designers, will also apply to these contractors with regard to the design element of their contracts.

The principal contractor

The principal contractor is generally responsible for the management of health and safety during the construction phase of a project. This will include co-ordination of all work with regard to health and safety. The Regulations impose duties on the principal contractor to ensure co-operation between all contractors, to develop and implement the health and safety plan, to ensure that the health and safety plan sets out the arrangements for health and safety management, to ensure that all contractors and employees comply with rules in the health and safety plan for the project, to ensure that only authorized persons are permitted on the site, to provide the planning supervisor with information for the health and safety file, to monitor health and safety performance, to provide opportunity for health and safety discussion with all persons at work on the project and to ensure that all contractors have the necessary information and training in respect of health and safety. The role is generally about ensuring an integrated approach to health and safety on site.

The Regulations give the principal contractor the right to give directions to any contractor to enable compliance with these duties. The principal contractor can also lay down rules, for the purpose of health and safety, for the management of construction work within the health and safety plan. The principal contractor must display on site (Regulation 16(1)(d)) the information provided in the notification to the Health and Safety Executive.

The risk assessments prepared by other contractors should be evaluated by the principal contractor. The objective of the evaluation should be to ensure that all risks are considered in the overall management of health and safety and that a co-ordinated approach to risk management, which takes into account the inter-relationship of different contractors' operations and their associated risks, is adopted. The principal contractor must be satisfied that the risk assessments and method statements prepared by contractors are adequate and compatible with the health and safety plan.

Communication is essential in ensuring that the health and safety plan is effectively implemented. The principal contractor must clearly communicate the requirements of the health and safety plan and ensure that arrangements for health and safety are coherently disseminated throughout the site on a regular basis.

Where a principal contractor is involved in the selection and appointment of works contractors or sub-contractors, the principal contractor should have regard to ensuring that these contractors have sufficient information regarding the health and safety requirements of the project and that they have allowed sufficient and appropriate resources in their tenders.

Contractors

The principal contractor's co-ordination role does not relieve individual contractors of their legal duties. The Regulations impose requirements and prohibitions on *all* contractors. They are required to co-operate with the principal contractor, provide information to the principal contractor which might affect the health and safety of persons at work on the project, comply with directions given by the principal contractor under the Regulations, comply with rules in the health and safety plan, provide information in accordance with the Reporting of Injuries, Diseases and Dangerous Occurrences Regulations 1995, and provide the principal contractor with information pertinent to the health and safety file. Contractors should not permit employees to work on site unless they have the name of the planning supervisor, the name of the principal contractor, and relevant parts of the health and safety plan.

The Management of Health and Safety at Work Regulations 1999 require Employers to assess the risks to health and safety arising out of their undertakings. Contractors are required to provide these risk assessments, together with details of the arrangements for health and safety resulting from the risk assessments, to the principal contractor. This enables the latter to develop and co-ordinate an integrated approach to health and safety management for a project.

Tender procedures

Effective contractor selection is very important in the management process. Tender documentation should be set out to enable judgements to be made about the adequacy of resources allowed in tenders and to prompt contractors to provide information confirming their competency. Documentation should contain sufficient information to make contractors aware of the health and safety requirements for a project and thus enable them to make adequate financial and programme provision. Tender pre-qualification procedures will help to ensure that only competent contractors are invited to tender (see Chapter 4).

Risk assessment

Risk assessment is at the heart of health and safety management. The purpose of risk assessment is to help designers to make informed design decisions. It is about identifying hazards arising from a design solution, assessing the likelihood that harm will occur from the hazard and the probable severity of the harm caused by a hazard occurring. Common sense dictates that priority should be given to managing those risks where the severity of harm (the consequence) and the likelihood of occurrence (the frequency) are high. An approach to risk assessment is shown in Figure 15.2.

Hazard identification involves a systematic analysis and review of the design. Identification can be guided by a knowledge of published accident statistics (such as those published by the HSC), which will help identify where and how construction accidents are

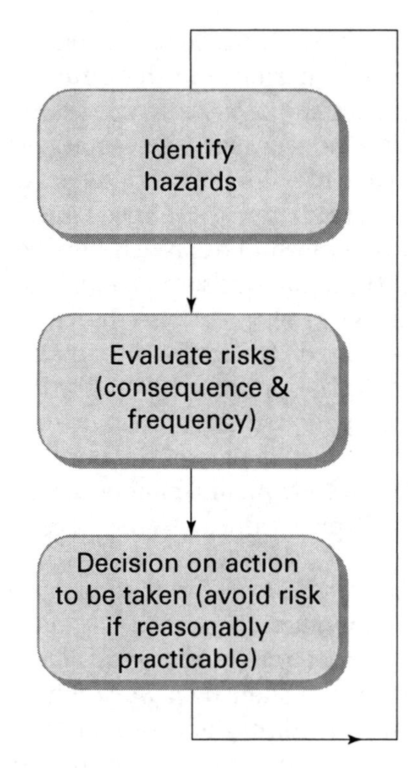

Figure 15.2 Approach to risk assessment

likely to occur. Risk evaluation can be undertaken using a combination of qualitative and quantitative methods.

Even crude assessments of risk are useful in aiding decision-making, and personal knowledge, experience and 'gut feeling' should not be ignored when evaluating risk. The risk assessments are essential in making rational decisions about the design and in highlighting those areas of the design where contractors need more information regarding health and safety issues.

Competence and resources

There is a duty on clients to appoint persons as planning supervisors only if they are reasonably satisfied that they are competent to perform the functions required of planning supervisors under the Regulations for the project and that they have, or will allocate, sufficient resources for the project to comply with the Regulations. The Regulations also impose a duty on any person arranging for a designer to design or a contractor to construct to be reasonably satisfied of competence and resources in performing these functions in accordance with the Regulations for the project. To ensure reasonable satisfaction, some checks will be necessary. For small or low-risk projects, it may be a simple case of asking questions on a face-to-face basis to ensure adequacy of competence and resources. For larger or more complex projects, a more formal approach may be appropriate – for example, giving written questions and requiring both written and oral responses. The types of questions that should be asked will depend on the particulars of the project, but in all cases they must be designed to obtain information that will enable the assessor to make a judgement about the adequacy of competence and resources. These questions could be included as part of the normal pre-selection procedures associated with the appointment of consultants, contractors and sub-contractors. Typical information requested might include:

1. Membership of relevant professional associations.
2. Knowledge of construction related to a particular project.
3. Knowledge of health and safety issues and requirements.
4. Ability to co-ordinate their own work with that of others.
5. Availability and suitability of human resources.
6. The time allowed to perform the required functions.
7. The availability and suitability of technical facilities.
8. The management systems that will be used.
9. The proposed methods of dealing with risks identified in the health and safety plan.
10. Arrangements for managing health and safety.
11. Arrangements for monitoring compliance with health and safety legislation.

The Regulations in respect of competence and adequate resources require reasonable satisfaction at the time of appointment or arrangement. There is no requirement for competence and resources to be continually monitored once appointments or arrangements have been made.

Relationship of the Regulations with other health and safety legislation

The Regulations apply simultaneously with other health and safety legislation. Many of the health and safety regulations relating to construction are specific in nature. They prescribe actions to be taken or provisions to be made to control or avoid specific risks. The Health and Safety at Work, etc., Act improved on this situation. The Act gave emphasis to individuals and their duties. The Management of Health and Safety at Work Regulations were first introduced in 1992 to improve health and safety management and to make more explicit the requirements of the Health and Safety at Work, etc., Act insofar as they place duties on Employers. They are applicable to all Employers and they apply to all work on construction sites.

The Management of Health and Safety at Work Regulations 1999 generally require Employers to take an active role in managing health and safety in the workplace. Employers are required to assess risks, make appropriate arrangements for implementing measures in the light of risk assessment, provide health surveillance for employees, appoint competent persons to help devise and apply health and safety measures, develop emergency procedures, provide information and ensure adequate training for employees on health and safety, co-operate and co-ordinate activities with other Employers sharing the same workplace, and provide temporary workers with some health and safety information.

The Construction (Design and Management) Regulations 1994 extend the duties imposed by the Management of Health and Safety at Work Regulations. For example, designers have a duty under the Management of Health and Safety at Work Regulations to assess the risks arising out of their business; the 1994 Construction Regulations explicitly extend this duty to considering the risks in constructing a design. The Construction (Design and Management) Regulations address the specific characteristics involved in managing safety in construction. They take account of the construction process, the parties involved in the process, the complex contractual arrangements and the variety of different procurement and employment arrangements.

References

1. HSE, *Managing Health and Safety in Construction: Construction (Design and Management) Regulations 1994: Approved Code of Practice* HSG 224 (Sudbury: HSE Books, 1995).
2. HSE, *Blackspot Construction* (London: HMSO, 1988).
3. JOYCE, R., *The CDM Regulations Explained* (London: Thomas Telford Publications, 1995).

Bibliography

1. HSE, *Managing Health and Safety in Construction: Construction (Design and Management) Regulations 1994: Approved Code of Practice* HSG 224 (Sudbury: HSE Books, 1995).
2. JOYCE, R., *The CDM Regulations Explained* (London: Thomas Telford Publications, 1995).
3. BARRETT, B. and HOWELLS, R., *Occupational Health and Safety Law*, 2nd edition (London: Pitman Publishing, 1995).
4. HSE, *Designing for Health and Safety in Construction* (Suffolk: HSE Books, 1995).
5. HSE, *A Guide to Managing Health and Safety in Construction* (Suffolk: HSE Books, 1995).
6. The Institution of Civil Engineers, *The Management of Health and Safety in Civil Engineering* (London: Thomas Telford Publications, 1995).

16

Collateral warranties

Introduction

Collateral warranties create contractual agreements between parties where they would otherwise not exist. The use of collateral warranties on construction projects rose dramatically during the 1990s. This increase may be attributed to a variety of factors, including

1 Developments in the law of tort.
2 Changes in building procurement practice.
3 The ever-increasing complexity of financial arrangements and investment interests associated with development projects.
4 The increase in multiple ownership of development projects.

As a result of this increase, it is now essential for surveyors to have a basic understanding of the law and practice relating to the use of collateral warranties. Surveyors may be required to provide collateral warranties in respect of the services they are offering to the benefit of third parties, and so a clear understanding of the implications of such agreements is essential. Surveyors may be required to give advice to their clients with regard to obtaining collateral warranties for a development project, so, again, an understanding of collateral warranties is necessary. As Pike[1] notes, 'it is incumbent upon design professionals and other advisers of property investors to ensure that appropriate collateral warranties are obtained. If they do not, they may be held liable to their clients for negligence'.

This chapter defines collateral warranties, explains why they may be required, considers who may require them and from whom and discusses some of the common provisions contained in them.

General principles

Consider the situation where a leaseholder has just leased a building from a developer on the basis of a full repairing and insuring lease. On discovering design defects, what action can the leaseholder take? The leaseholder, while being responsible to the developer for rectifying the defects, would want to recover the cost from the negligent architect. But, in the absence of a collateral warranty, there is no contractual relationship between the leaseholder and the architect. Prior to 1990, it was assumed that there existed a remedy for the leaseholder in tort. However, subsequent cases, in particular, the decision in *Murphy v Brentwood District Council [1990] 2 All ER 908*, have restricted liability in tort and created some uncertainty about rights in tort. The response to this has been an increase in demands for collateral warranties from all those who may have an interest in a development, thus providing a contractual remedy in such situations.

Consider the situation of a contractor's insolvency during the course of a project. Does the Employer have the right to the benefits of sub-contract agreements to enable completion of the project with minimum disruption? In the absence of contractual relationships with the sub-contractors (which could be provided by collateral warranties), the answer is No.

Collateral warranties create a direct contractual link between parties where one would otherwise not exist. They are agreements alongside (collateral to) another contract. So, typically, for a surveyor, a collateral warranty will be collateral to the terms of engagement. For a contractor, a collateral warranty will be collateral to the building contract. For a sub-contractor, a collateral warranty will be collateral to the sub-contract agreement. Collateral warranties give contractual remedies to parties who, as a consequence of the doctrine of privity of contract, would not otherwise have them.

The presentation of collateral warranties can vary from standard forms of agreement to simple letters from the warrantor to the warrantee. The names used to describe these agreements also vary; collateral warranties are sometimes referred to as duty of care deeds, deeds of responsibility, etc. However, for a collateral warranty to exist, the normal prerequisites for a contract (offer and acceptance, consideration and the intention to create legal relations) must be observed.

Warranties can be created by either

(i) a simple contract requiring consideration; or
(ii) as a deed, in which case consideration is not required, but there are some formalities to be performed with regard to the execution of the deed; or
(iii) occasionally, it may be possible that a collateral warranty is implied as a result of advice and representations of a party (*Shanklin Pier v Detel Products Ltd (1951) 2 Ll Rep 187*).

In the absence of specific limitation wording, the effective difference between a simple contract and a deed is the limitation period. With a simple contract, the time limit for pursuing a claim is 6 years from the date of breach. For a deed, the period is 12 years. Thus it would be inappropriate to have a collateral warranty created by deed running alongside a main agreement with a limitation period of 6 years, as the warranty would be more onerous than the main agreement.

It is the view of many legal experts that the liabilities created by collateral warranties are greater than those that exist in tort. For example, designers who demonstrate reasonable skill and care would not be considered negligent in tort, whereas with a collateral warranty the standards required may, depending on the specific wording used in the warranty, be higher than this. In tort, with the exception of reliance duties, economic losses (for example, trading losses) are not recoverable, but under a collateral warranty, in the absence of an express limitation provision, economic losses are recoverable; the liabilities are therefore potentially greater.

Interested parties

Any individual or organization with a financial interest in a development project, but without a direct contractual relationship with the designers or builders, may seek the protection afforded by collateral warranties from these parties.

Consultants, contractors, sub-contractors and suppliers all have the potential to cause loss by their errors. Under certain circumstances, all may be required to provide a warranty to the benefit of a third party. The need for, and the format of, warranties will vary, depending on the nature of the project, the contractual relationships and the procurement strategy adopted.

Under traditional forms of procurement, contractors are responsible for the standard and quality of workmanship and materials and for the completion of work within a prescribed time period. This responsibility extends to include the work of sub-contractors and materials from suppliers. Collateral warranties may be required from contractors by funding institutions (who have an interest in the successful completion of the project and may want the right to act as Employer in, for example, the event of developer insolvency), by future tenants (particularly those tenants with a full repairing and insuring lease) and by future purchasers. Where a contractor has passed on sections of design work to a sub-contractor or supplier, the responsibility (to the Employer) for this design will still rest with the contractor. However, if the contractor were to go into liquidation during or after the completion of the works the Employer would have no contractual link with the sub-contractor or supplier and therefore no redress under the contract for any defective design work they may have carried out. Therefore in such instances it would be advisable for the Employer (and, indeed, any other interested party) to obtain a collateral warranty from the sub-contractor or supplier in respect of the design work.

Consultants generally have responsibilities for design and management. For a typical commercial development, consultants will be appointed by a developer with whom they will be in direct contract. Again, warranties may be required by funding institutions, future tenants and future purchasers of the development.

Figure 16.1 shows the typical contractual arrangements for a development project procured using the traditional approach. Some of the possible situations where collateral warranties may be used or required are also shown in Figure 16.1. If there are many tenants or purchasers, or if there are many funders with an interest in the project, then a large number of separate collateral warranties may be required; managing warranties could then become complex (and expensive).

Under other forms of procurement, the situation regarding warranties will vary. When advising clients on the need for protection

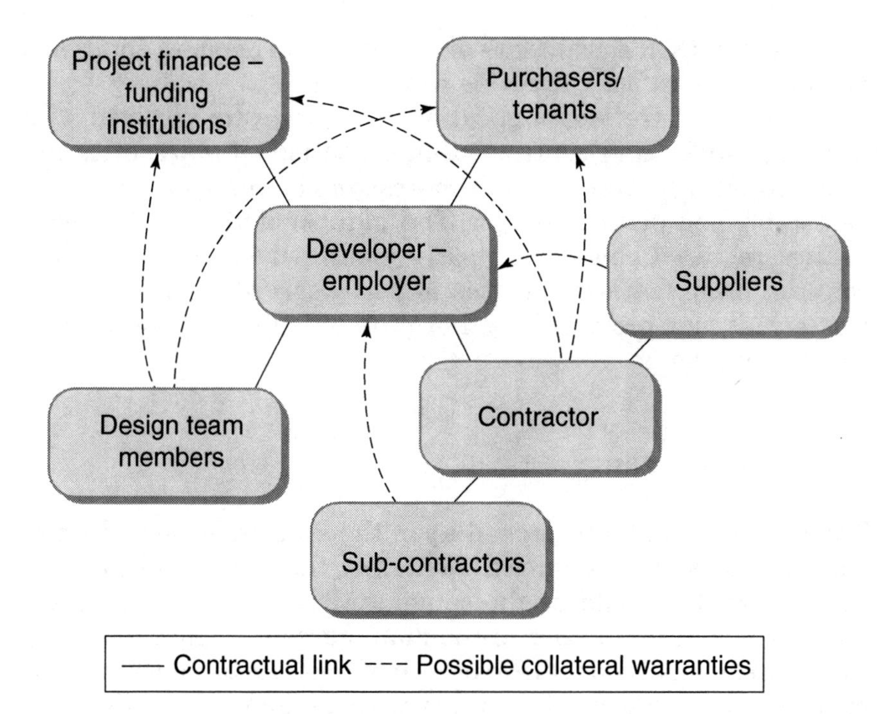

Figure 16.1 Contractual arrangement for traditional procurement

by way of warranties, surveyors should adopt a systematic approach. They should identify the likely causes of loss and who might be responsible for the loss arising. They should then examine the contractual arrangements to identify the protection afforded without collateral warranties, and where insufficient protection is available they should consider using them. As noted by Kitchen,[2] 'a collateral warranty is simply a way to shift responsibility for certain losses from one party to another'.

If collateral warranties will be required, it should be so stated in the main agreement. For example, if consultants are required to provide warranties for funders, tenants and purchasers, it should be stated in their terms of engagement. The information included in the main agreement should be such that the wording of the warranties and the beneficiaries are stated. This will enable the consultants to include in their fees for the costs, in terms of additional administration, professional indemnity insurance costs, etc., of providing the warranties. If insufficient, or no, provision for providing warranties is made within the original terms of engagement, then consultants are at liberty to demand additional payment or to refuse to provide warranties.

There are, therefore, implications for surveyors involved with drafting tender and contract documentation. If warranties are required from contractors, sub-contractors or suppliers, it should be established prior to tender. The number and format of warranties required should be clearly identified in the tender and contract documentation. If this is not observed, then obtaining warranties may prove to be either prohibitively costly or difficult (they may even be refused).

Terms of warranty

The terms of warranty depend upon the specific nature of a project and the benefits required. Generally, the obligations imposed by a warranty should be the same as those in the main agreement; it is normal to refer to the main agreement in a warranty. Therefore, if an architect is in breach of contract to a developer, he could also be in breach of warranty with a tenant. The obligations imposed by a warranty should be no greater than those imposed by the main agreement. Also, care should be taken to ensure that there are no terms included in warranties that are in conflict with terms in the main agreement. Although the format and composition of warranties varies enormously, the following are areas commonly dealt with:

1 *Limitation of damages* – in view of the potentially greater liability in contract than tort, it is common to restrict the damages recoverable, e.g., covering the cost of remedying defects only.
2 Use of reasonable skill and care in constructing or designing.
3 Design or construction fit for its intended purpose.
4 *Insurance* – for example, consultants may be required to take out and maintain professional indemnity insurance. Warrantors should ensure that they inform their insurance companies with regard to all collateral warranties they provide to ensure that their insurance cover is valid.
5 *Assignment* – occurs where the rights or obligations of one of the parties to a collateral warranty are transferred to a third party, for example, a second tenant. The law associated with the assignment of warranties without consent is complex. In this

respect, it is appropriate to make express provision within a warranty with regard to assignment; it should state whether the warranty can be assigned and, if so, how, to whom and how many times.

6 *Limitation* – to ensure that a warranty is no more onerous than the main agreement to which it relates, it is sometimes advisable to include a clause limiting the validity of the warranty from a fixed date, for example, the date of the main agreement, for a specified period of time.
7 *Novation rights* – for example, the right of an Employer to take over a sub-contractor's appointment in the event of a breach of contract or sub-contract by the main contractor.
8 *Preservation of rights at common law* – a specific clause is common.

In situations where joint liability could occur (for example, bad workmanship by a contractor coupled with negligence on the part of the contract administrator in the supervision of the Works), warrantors should ensure that warranties are provided by all those with whom they may share joint liability. If this is not ensured, the total liability to the warrantee will be borne by the single warrantor.

Any person or organization that is either bound by a warranty or entitled to enforce a warranty must be a party to that warranty. Thus, with a warranty for architectural design to the benefit of a funder, the parties to the warranty will be the architect (the warrantor), the funder (the warrantee) and the developer (as the funder will want the right to take over the architect's appointment in the event of breach by the developer, the developer will need to be a party to the warranty to grant this right).

There is a large, and increasing, number of standard forms of warranty available for use on construction projects. Many of these have been drafted by trade, professional and client organizations (for example, the Royal Institute of British Architects, the British Property Federation, etc.) to protect the interests of their members. The JCT has produced a number of standard collateral warranties to complement their suite of contracts; a few examples are identified below

- Contractor collateral warranty for a funder (CWa/F)
- Contractor collateral warranty for a purchaser or tenant (CWa/P&T)

- Sub-contractor collateral warranty for a funder (SCWa/F)
- Sub-contractor collateral warranty for a purchaser or tenant (SCWa/P&T)
- Construction management trade contractor collateral warranty for a funder (CMWa/F)
- Construction management trade contractor collateral warranty for a purchaser or tenant (CMWa/P&T)
- Management works contractor/employer agreement (MCWK/C).

Additionally, the use of bespoke warranties drafted by solicitors to cater for the interests of their clients is widespread. In all situations, whether responding to a request to provide a warranty or advising on the use of warranties, the obligations imposed by warranties should be carefully considered to ensure that they do not impose unrealistic obligations on the warrantor or fail to provide the benefits required of them.

Contracts (Rights of Third Parties) Act 1999

For many years the English legal system supported the principle of 'privity of contract', which means that it is only the parties who have entered into a contract who have a right to any of the contractual remedies contained within that agreement. It is because of this principle of privity that the use of collateral warranties has become so commonplace, i.e., they create a legal relationship where one would not normally exist. But as explained previously on large projects where a considerable number of funders, purchasers and tenants may be involved, the use of collateral warranties can produce a mountain of legal paperwork, a situation that may be remedied through the use of the Contracts (Rights of Third Parties) Act. The main principle of the Act is to allow a third party (i.e., a party who has not entered into the contract) the legal right to enforce a contractual term that confers a benefit upon the third party. It is important to note that the Act only allows a third party to enforce conferred benefits; it is not possible for an obligation to be imposed on a third party through this Act. The JCT first made use of this relatively new legislation in the first edition of the Major Project Form (2003). The contract includes two standard collateral warranties for the benefit of funders, purchasers and tenants. The contractor will have agreed to the use of these warranties when entering into the contract and as a result funders, purchasers and tenants, as third parties to the contract, are able to use the Act to enforce their benefits under the warranties. Therefore the use of one document, the Major Project Form, provides warranties to any number of funders, purchasers or tenants associated with the project.

Conferring benefits under the JCT form

In the Standard Form with Quantities, the JCT has provided three methods by which benefits may be transferred to third parties.

- Assignment
- Third party rights
- Collateral warranty.

Assignment

A simple explanation of assignment is where a party to a contract assigns (i.e., transfers or makes over) a benefit or benefits from a contract to a third party. The JCT initially prevents either the Employer or contractor from assigning any rights under the contract unless they have a written agreement from the other party (see clause 7.1). Either party may refuse to give their consent to an assignment and there is no requirement that any refusal must be reasonable.

The limitations placed upon assignment in clause 7.1 may be modified by the use of the optional clause 7.2, which allows the Employer to assign benefits to a third party where the freehold or leasehold interest of the whole works (or any section) is transferred to a third party after practical completion. The rights passed on would allow an assignee to commence legal proceedings (in the Employer's name) against the contractor to enforce a benefit originally enjoyed by the Employer. The legal proceedings may take the form of arbitration or litigation and would be dependent upon which method is allowed for in the original contract (see Articles 8 and 9). If the Employer wishes to use this right of assignment this must be notified in the Contract Particulars.

Third party rights

In the 2005 edition of the Standard Form of Contract the JCT opted to make use of the Contracts (Rights of Third Parties) Act, as previously described. If an Employer wishes to make use of this legislation, the appropriate details and information must be made available, or referred to, in Part 2 of the Contract Particulars. Under this procedure third party rights may be made available from the contractor to an identified purchaser, tenant or funder. It is obviously not possible to obtain third party rights from a sub-contractor via the main contract; this may only be achieved by inserting the appropriate terms into the sub-contract. A contractor's liabilities and a third party's benefits are detailed in Schedule 5 of the contract where Part 1 is relevant to purchasers and tenants while Part 2 is relevant to a funder.

By adopting this procedure the Employer is able to reduce the volume of paperwork compared to using individual collateral warranties. For example the project detailed in Appendix A comprises five shop units and ten flats and in the normal course may require the preparation and signing of fifteen collateral warranties for purchasers and tenants and possibly one or two warranties for funders. By using the third party rights the need for this extensive paperwork is removed as long as the parties have been identified in Part 2, either by name, class or description.

Collateral warranties

The JCT has still retained a provision for using collateral warranties to allow third parties to acquire rights against the contractor and his sub-contractors. In Part 2 of the Contract Particulars where third parties are identified, it is necessary to specify how their rights are to be conferred, i.e., refer to clause 7A or 7B where third party legislation is to be used, and refer to clause 7C or 7D where collateral warranties will be required. It is possible to have a mixture of the two, i.e., some parties gaining rights under the Contracts (Rights of Third Parties) Act and others being in possession of collateral warranties. The collateral warranties referred to in clauses 7C and 7D are the standard forms produced by the JCT for use by purchasers, tenants and funders. A contractor is required to enter into the relevant collateral warranty within 14 days of receiving a notice from the Employer.

Sub-contractor warranties

The contractor may be informed, via the main contract, that collateral warranties will be required from his sub-contractors. An Employer is advised to be selective in this requirement[3] and limit its use to sub-contractors who are responsible for design works or for key aspects of the project. At the contract stage it is possible that the names of the relevant sub-contractors are not yet known, in which case they would have to be identified by other means, i.e., by type or category of work. Having been informed of the Employer's requirements a contractor is now aware that he must inform any relevant sub-contractors of the need for a collateral warranty during sub-contract negotiations. The collateral warranties are to be the JCT standard forms for use between a sub-contractor and purchaser, tenant or funder. In addition the Employer may require the provision of an Employer/sub-contractor warranty. The JCT has stated its intention to produce such a standard form of warranty[4] but until one is published Employers will have to commission a bespoke warranty.

A contractor is to obtain an appropriate warranty from his sub-contractor within 21 days of a notice form the Employer. It is permitted for a sub-contractor to suggest amendments to any of the warranties. Both the Employer and contractor must approve such amendments, but they cannot unreasonably delay or withhold their agreement.

Liability to third parties

When granting rights or a warranty, a contractor's liability to a purchaser and tenant is not the same as his liability to a funder.

Purchaser and tenant liability

From practical completion a contractor is liable to pay the reasonable costs of a purchaser or tenant that have been incurred as a result of defective work. Through optional paragraph 1.1.2 a contractor may also be liable for consequential costs, i.e., loss of profits or cost of alternative accommodation, up to a financial limit stated in the Contract Particulars. A net contribution clause exists which limits the contractor's liability to only those costs that can be fairly shown to be his responsibility – it is the purchasers and tenants responsibility to recover any remaining costs from the other responsible parties, i.e., consultants and/or sub-contractors.

These rights or warranty may be assigned to subsequent purchasers or tenants up to a maximum of two occasions. It is not necessary to obtain the contractor's consent to the assignment but it will only become effective when the contractor is given a written notice of the event.

Funder

The contractor's liability to a funder commences upon the signing of the contract (for third party rights) or the signing of the collateral warranty. The contractor is liable to the funder for costs resulting from the contractor's breach of the main contract. The contractor's financial liability is limited by a net contributions clause. The rights or warranty may be assigned up to a maximum of two occasions.

The granting of rights or a warranty to a funder can be beneficial to a contractor. If the Employer acts in a manner that would allow the contractor to terminate his employment or if the funder terminates the finance agreement with the Employer, it is possible for the funder to step in and take the place of the Employer. This procedure would allow the contractor to stay on site and complete the works, which in most cases would be far more beneficial to the contractor than exercising his right of termination against the Employer.

Period of liability to purchasers, tenants and funders

With reference to the granting of rights, the contractor's period of liability will be determined by the manner in which the main contract was executed. Where the contract was executed under hand the contractor's liability extends for 6 years from the date of practical completion of the Works on any section, where relevant. The period is 12 years where the contract was executed as a deed.

A similar situation exists with warranties. Where a contract is executed as a deed then any warranties must also be executed as a deed. Where the contract is executed under hand the JCT advises that any warranties may be executed under hand, which seems to imply that an Employer could require the warranty to be executed as a deed (see clause 7.5).

An alternative to collateral warranties

For problems occurring after practical completion of a project, a form of insurance cover, often referred to as Building Users' Insurance against Latent Defects (BUILD),[5] is becoming popular with developers. As Paterson[6] notes, this insurance 'is a non-cancellable material damage policy for the benefit of the developer, subsequent owners and occupiers'. The insurance deals with many of the concerns of tenants and purchasers for which warranties are used, but as it is only effective from practical completion, it does not deal with all the relevant issues.

Such policies normally cover the shell and structural members (e.g., foundations and frame, external walls and claddings, roofing, floors, stairs and any other structural component) against 'inherent defects'. A noticeable number of items are excluded from such a list, including heating and ventilation installations, lifts, escalators and electrical distribution systems. Insurers are prepared to provide cover for some of these items through add-on options, but this will obviously increase the premium to be paid by the Employer.

The cover is normally given subject to the appointment of an independent consulting engineer, who will monitor the project on behalf of the insurer, therefore the insurance can only be applied to 'new build' projects, where the engineer may oversee the work. The policy may not be issued retrospectively. At the time of issue of the certificate of practical completion, the consulting engineer will also issue a certificate of approval for the works. The certificate may contain certain qualifications regarding the structure where the engineer does not approve the detail but the Employer and his advisers have decided to retain it, and if so, these items will be excluded from the policy. The policy is then issued; it is non-cancellable and will normally run for a period of 10 years. The cover provided will normally cover the cost of repairing property damaged as a result of a latent defect. This would include the cost of repairing the damaged element and any consequential damage it caused to the building and fabric; also the cost of removing debris, etc., and professional fees.

It is possible for the Employer to assign the policy to a subsequent purchaser or, if the property is let, to have the tenants' names noted on the policy to allow them to enjoy the same cover. This obviously has potential benefits to the Employer, prospective purchasers and tenants. However, nothing comes free, and the Employer will obviously incur a considerable cost when taking out such a policy, e.g., the cost of a basic policy and independent engineer's fee could amount from 1% to 2.5% of the project sum.

From a contractor's or consultant's point of view, there may still remain some liability in respect of the insurers' rights of subrogation with BUILD unless either the contractor and consultant are named parties in the insurance policy or subrogation rights are waived by the insurers, but again at an additional cost to the Employer.

References

1. PIKE, A., Collateral Warranties under construction and civil engineering contracts, *Architect and Surveyor* (March 1990), p. 19.
2. KITCHEN, S., Paper Mate, *Building* (31 January, 1992), p. 20.
3. JCT, Standard *Building Contract With Quantities* (Sweet & Maxwell, 2005), Footnote [31] and see guide.
4. JCT, *Standard Building Contract Guide* (Sweet & Maxwell, 2005), note 93.
5. Building EDC, *Build Report* (London: NEDO, 1988).
6. PATERSON, F. A., *Collateral Warranties Explained* (London: RIBA Publications, 1991).

Bibliography

1. PATERSON, F. A., *Collateral Warranties Explained* (London: RIBA Publications, 1991).
2. WINWARD, Fearon, Collateral Warranties – A Practical Guide for the Construction Industry (London: BSP, 1990).
3. KITCHEN, S., Paper Mate, *Building*, 31 January 1992, pp. 20–21.
4. CHAPPELL, D., Collateral Warranties – RIBA and RIAS Forms, *Architects Journal*, 26 July 1989, pp. 67–69.
5. PIKE, A., Collateral Warranties under construction and civil engineering contracts, *Architect and Surveyor*, March 1990, pp. 19–21.
6. CLARKE, R., Collateral Warranties – the current state, *Chartered Quantity Surveyor*, April 1990, pp. 9–11.

17

Guarantees and bonds

Introduction

The costs to an Employer following the failure of a contractor during the course of a project can be significant. Similarly, in today's industry, a contractor's success is often dependent on the performance of sub-contractors. The failure of a sub-contractor can mean a potentially profit-making contract resulting in loss for a contractor. The objectives of guarantees and bonds are to encourage performance and to provide financial recompense in the event of failure.

This chapter examines the purpose of guarantees and bonds, discusses their various sources and formats, considers the arguments for and against their use, and discusses situations where their use may be appropriate.

The purpose of guarantees and bonds

Guarantees and bonds are arrangements whereby the obligations that one party owes to another under an agreement are guaranteed by a third party. The third party, usually an insurance company or a bank, promises to pay a sum of money in the event of non-performance. The following types of bonds and guarantees are often used on construction projects.

Performance bonds/guarantees

These are, as the name implies, aimed at ensuring the satisfactory performance of a party to a contract. They are the most commonly used type of bond/guarantee used in the construction industry. They may be required by an Employer to ensure the performance of a contractor, or by a contractor to ensure the performance of a sub-contractor. In the event of non-performance, the guarantor or bondsman provides financial compensation. The usual limit of compensation is 10% of the contract sum.

Tender bonds/guarantees

These are obtained to ensure that a successful tenderer will enter into a contract. Their value varies, but is usually somewhere between 1% and 5% of the tender sum. One of the purposes of tender bonds/guarantees is to provide financial recompense for the costs associated with the refusal by a tenderer to honour a tender. Most typically, they are requested by a contractor to ensure a sub-contractor's commitment to enter into a sub-contract agreement. The associated costs and disruption to a contractor's programme in the event of a sub-contractor withdrawing an offer can be significant. A new sub-contractor will have to be found and appointed, which may cause delays to the programme resulting in liquidated damages becoming payable to an Employer and/or acceleration costs for the completion of the project. It may also not be possible to obtain such a competitive tender the second time around (and, probably, at short notice), and thus the project costs will increase, leading to a reduction in profit. A tender bond/guarantee may discourage withdrawal by a sub-contractor, or at least provide some financial recompense to a contractor in the event of withdrawal.

Payment bonds/guarantees

These ensure the obligation to make payment. They may be requested by contractors to guarantee an Employer's duty to make payments or by sub-contractors to ensure payment from a Contractor.

In most situations, the protection afforded by guarantees and bonds will be required in the event of insolvency. The effects of insolvency on construction projects are discussed in detail in Chapter 19.

Sources and formats of bonds

The two most commonly used sources of protection are bank guarantees and surety bonds. The terms 'guarantee' and 'bond' are used synonymously within the industry. There is, however, a difference in the definition of the two terms. McDevitt[1] draws the distinction between demand guarantees and performance bonds. To summarize:

Demand guarantees are documentary covenants made by a guarantor to indemnify a beneficiary, subject to the conditions in the covenant. The guarantee is an agreement between the guarantor and the beneficiary. Thus, if an Employer is given a demand guarantee by a bank in respect of the obligations of a contractor, the contractor is not a party to the guarantee agreement. The beneficiary is therefore in a strong position should there be a default. Demand guarantees are contracts and can be created by either a simple contract or executed as a deed.

Demand guarantees are usually given by banks. There are two basic types: on demand guarantees (often referred to as on demand bonds) and documentary demand guarantees. On demand guarantees basically require a guarantor to make payment to a beneficiary upon request to do so. In the case of documentary demand guarantees, payment will only be made on the furnishing, by the beneficiary, of the documents required by the terms of the guarantee. These, for example, may be documents proving a court judgement.

Banks favour demand guarantees because they do not need to get involved in legal arguments and disputes following a default; their position is generally straightforward. However, the situation is not so satisfactory for those required to provide demand guarantees. Take, for example, a contractor required to provide a 10% demand guarantee in respect of a £2,000,000 contract. The guarantee will be for the sum of £200,000. The contractor's bank issuing the guarantee will treat the value of the guarantee as contractor's credit and will, therefore, reduce any credit facilities offered to the contractor by this sum. In addition, the bank will probably require security from the contractor to support the credit. Both these actions will affect a contractor's cash flow and make it more difficult for him to perform contracts. Indeed, the operating capacity of a construction firm can be reduced by the requirement to provide demand guarantees. A contractor in this position may also feel exposed, especially where on demand guarantees are provided. The contractor has minimal rights to prevent a bank paying against an on demand guarantee. Banks will pay on demand and leave the contractor to settle any dispute directly with the beneficiary.

Performance bonds are three-party agreements between a surety (or bondsman or guarantor), a beneficiary and a principal debtor.

In essence, the surety guarantees the performance of the principal debtor. In the event of a default by the principal debtor, the surety will make good the loss caused to the beneficiary up to the financial limit of the bond. Performance bonds will usually be provided jointly and severally in the name of the principal debtor and the surety; the surety would, therefore, rely first on the principal debtor satisfying any claim before becoming involved. The principal debtor receives no protection from the surety under the bond; it is not an insurance policy. Thus, the only situation where a surety is likely to be required to make payment would be in the event of the insolvency of the principal debtor.

Performance bonds are usually issued by insurance or surety companies. Unlike banks, these organizations will have little information regarding the financial standing of a principal debtor. Surety companies will, therefore, usually undertake extensive checks to ascertain the risk associated with providing a bond. The cost of the bond, together with the knowledge that an organization is able to obtain a bond, will often provide some comfort to a beneficiary that a principal debtor is sound.

Unlike demand guarantees, the beneficiary of a performance bond will be required by the surety to prove loss before payment is made. Additionally, the bond will not affect a principal debtor's credit facility, nor will it usually be necessary to provide security. As with demand guarantees, bonds can be created by simple contract or deed.

Performance bonds are often required by Employers in respect of a contractor's obligations under a construction contract. Similar protection may be required by contractors from their sub-contractors.

The arguments for and against bonds and guarantees

If good practice is observed in the selection and appointment of contractors and sub-contractors, it could be argued that there is no need to obtain bonds and guarantees. Certainly, the requirement to provide protection in the form of guarantees or bonds will, in most cases, add to the cost of a project. Where a demand guarantee is required from a contractor's bank, the resulting cash flow problems have been discussed above.

In considering the need for bonds, a sensible approach is required. While the comfort provided to Employers by a guarantee or bond may be reassuring, the costs of providing the protection may not be worth the benefits obtained. Consider the situation where an Employer is a frequent developer or has a substantial development programme; the cost of obtaining performance bonds or guarantees for all contracts would probably be high compared to the value of claims resulting from those bonds. It may be more appropriate, therefore, for the Employer in this situation to accept the risk of contractor default rather than require a large number of expensive bonds.

The use of bonds

There is no definitive list of situations where a bond should be used, and a decision will need to be made for each project. In some circumstances, a client may insist that a contractor provides a performance bond; in others, the design team may need to advise a client on the availability, suitability and cost of the various forms of protection available. Situations where a bond may be particularly appropriate would include:

(i) to protect the interests of a 'one-off' developer;
(ii) where a new or unproved contractor is involved in a project;
(iii) tender bonds in respect of nominated sub-contracts;
(iv) where a bond is thought appropriate for the risks inherent in a project.

If a bond will be required, it should be clearly stated in any tender and contract documentation. The terms of the bond should include the financial limit, details regarding the release of the bond, and defaults and non-performance covered by it. Care should be taken to ensure that the conditions of the bond are observed; for example, giving notices of default to the surety. Failure to do this may render a claim invalid. The alteration of contract terms will render a bond invalid (in the absence of express conditions in the bond permitting alteration).

Where a surveyor is requested to check the wording or form of a non-standard bond, specialist advice should be sought. Standard forms of bond available include those published by the ICE, FIDIC and ABI. The use of standard forms of bond is recommended. There is no JCT standard form of performance bond.

JCT bonds

The JCT has provided three bonds in Schedule 6 of the standard form of contract:

- Advance Payment Bond
- Bond for payment of off-site materials
- Retention bond.

Advance Payment Bond

Where the Employer has agreed to provide an advance payment(s) to the contractor it may be advisable to request the contractor to obtain an Advance Payment Bond. In fact the default position is that where the Contract Particulars show that an advance payment is to be made then an Advance Payment Bond is automatically required unless the option is deleted in the Particulars.

The Advance Payment Bond is a tripartite agreement between the Employer, contractor and a surety. The surety stands as a bondsman to provide a guarantee that if the Employer is not able to recover an advance payment from the contractor then the surety becomes liable to reimburse the Employer according to the detail of the bond.

The value of the advance payment(s) is inserted into the bond and this determines the maximum financial liability of the surety. The surety's liability starts on the date an advance payment is made to the contractor, and continues until the advance payment is repaid or until a 'longstop' date (inserted into the bond) has been reached. Whichever event occurs first signifies the end of the surety's liability.

If an Employer wishes to claim under this bond he must complete the Notice of Demand attached to the bond. The Employer's bankers must authenticate the Employer's signature or that of an authorized representative.

Bond in respect of payment for off site materials and/or goods

Within the standard form of contract the Employer may signify his willingness to pay the contractor for materials held off site. These materials are subsequently categorized as listed items which are uniquely identified or listed items which are not uniquely identified. Where the Employer is prepared to pay the contractor for such materials he has the option of requiring the contractor to obtain a bond for either category or both (see clauses 4.17.4 and 4.17.5). In either instance the form of the bond is the same, and is provided in Part 2, Schedule 6 of the contract.

The operation of the bond is similar to the Advance Payment Bond. The surety's liability is to reimburse the Employer for monies the Employer may have paid to the contractor for listed materials off site. The Employer is to insert a maximum value for which the surety may be liable. The surety's liability ends on the date all the listed items have been delivered to site or by a specified 'longstop' date; whichever occurs first.

JCT retention bond

Where the optional clause (4.19) is being used, the contractor is required to provide the Employer with a bond. A copy of the bond is contained within the contract in Schedule 6, Part 3. The bond, in favour of the Employer, is to be executed before or on the date of possession and be given to the Employer. The value of the bond is stated in the Contract Particulars and comprises a maximum aggregate sum. It is anticipated that this sum will be equal to the maximum amount of retention that would have been held by the Employer if the bond were not in existence. If the contract sum increases, because of variations or fluctuations, the original bond may not be sufficient to cover the amount of retention that would have been due. In this instance the contractor has the choice of increasing the value of the bond or allowing the employer to deduct retention from the monies not covered by the bond.

The bond is to be operative from the date of possession and is reduced by 50% after the issue of the certificate of practical completion (no specific mention is made of the effect of partial possession but the procedural rules do in fact take account of this).

As long as the surety is not in receipt of a written demand on the bond, then their liability under the bond ends after the earliest of the following events:

- the date of issue of the certificate of making good defects; or
- payment has been made against the bond up to the maximum aggregate amount; or
- the expiry date inserted in the contract appendix.

If the contractor fails to provide or maintain the retention bond, then the normal retention rules become operative and the Employer would deduct the due retention from the interim certificate immediately after the contractor's breach was discovered. If the contractor subsequently arranges the bond, the Employer is required to release the retention that had been withheld.

Where the optional clause 4.19 is being used, the provisions relating to retention in clauses 4.10.1 and 4.20 are not to apply. The exception to this rule is that a retention percentage is still to be inserted in the Contract Particulars and the architect, or quantity surveyor, is still required to prepare a statement of the contractor's retention to show what retention would be deducted if it were not for the bond. The reasons for this calculation are:

- to maintain a check that the nominal retention does not exceed the value of the bond;
- to allow the employer to inform the surety of the amount of retention that would have been held at the time a written demand is submitted.

The latter point is important because the surety's liability is not greater than the value of the nominal retention calculated at the time of the employer's demand against the bond. An employer must realize that, if a claim arises early on in the project, the value of the bond is very limited. The surety's liability gradually increases each month as it matches the nominal retention calculation, until the maximum aggregate sum is reached.

Claiming under the retention bond

If the employer wishes to make a demand on the bond, this must be submitted, in writing, to the surety. The signature of the employer must be authenticated by the employer's bank. When dealing with bonds it is important to ensure that all the required procedural detail is complied with. There have been instances in the past where sureties have tried to renege on their obligations by quoting breaches of procedure. In the written demand the employer is required to refer to the specific bond against which he is claiming, state the amount of nominal retention that would have been held at that time, and the amount that is being demanded. In the demand the employer is required to identify the reason for the claim, which must be for one of the events listed below:

- Costs have actually been incurred by the employer because of the contractor's failure to comply with a valid architect's instruction. This must be accompanied by a statement from the architect which confirms the contractor's failure.
- Insurance premiums, for building works, paid out by the employer, where the contractor has failed to maintain or take out the insurance for which he was responsible.
- Liquidated and ascertained damages for which the contractor is liable. This is to be accompanied by the architect's certificate confirming the contractor's failure to complete on time, issued under clause 2.31.
- Any expenses or direct loss and/or damage caused by the employer terminating the contractor's employment.
- Finally, a 'catch all' provision that refers to any other costs which the employer has actually incurred and is entitled to claim from the contractor under the contract. In this instance the employer has to clearly identify the contract provision that creates this entitlement.

To accompany the above demand, the employer must certify that the contractor has been given 14 days' written notice of his liability for the monies claimed, and that the contractor has not discharged this liability. At the same time the employer is required to send a copy of the notice to the surety. If all the above procedure is correctly followed, the surety will accept that the demand is conclusive evidence that the amounts claimed are properly due and payable to the employer by the contractor. However, this acceptance by the surety does not prevent the contractor from denying the breach. If the breach is disputed, the contractor could refer it to adjudication or proceed to arbitration or litigation. If the employer has also required the contractor to provide a performance bond, there may be instances where the employer could claim against either bond. If this situation arises, it is stated that the employer is to claim against the retention bond in the first instance.

Reference

1. MCDEVITT, K., *Contract Bonds and Guarantees*, The Chartered Institute of Building Occasional Paper No. 34 (1985), Berkshire.

Bibliography

1. MCDEVITT, K., *Contract Bonds and Guarantees*, The Chartered Institute of Building Occasional Paper No. 34 (1985), Berkshire.
2. POOLE, A., A clear view of the pitfalls, *Building*, 17 November 1995, p. 40.
3. RIDEOUT, G., Licence to kill, *Building*, 5 March 1993, pp. 20–21.
4. MINOGUE, A., Gentlemen don't prefer bonds, *Building*, 9 September 1994, p. 38.
5. SPENCER, K., No need to pay on demand, *Building*, 28 July 1995, p. 39.
6. NJCC, *Guidance Note 2 – Performance Bonds* (London: RIBA Publications, 1986).
7. Association of British Insurers, *ABI Model Form of Guarantee Bond – An Explanatory Guide*, September 1995.

18

Value management

Introduction

This chapter provides an insight into how the application of some of the techniques of value management can optimize the value generated from capital investment and operating expenditure associated with built assets. The chapter describes the evolution and history of value management and explains some of the processes, applications and benefits of the techniques. A selection of case studies is included to illustrate some of the issues and benefits outlined.

A brief history of value management

Today's managers are faced with the challenge of delivering more with less. Managers must continually make decisions on the best balance between conflicting factors. These factors include service levels, functional performance, resources, costs, time, environmental impact, social acceptability and so on. Best value is achieved through finding the optimum balance between these factors in the prevailing business and organizational environments.

Value management encompasses a range of management decision-making processes that enable key stakeholders to define and achieve their goals for achieving best value projects, products and services, using minimum resources, in a manner consistent with the broad goals of their business or organization.

Many of the principles of value management, as we know it today, were first introduced and formalized shortly after the Second World War by Larry Miles, an electrical engineer in the purchasing department of the General Electric Company. Substitute materials were identified to overcome shortages and thus ensure the continued production of some of GE's products. Miles found that this approach often led to improvements in the quality of the final product at a reduced cost – often contrary to expectation. Miles developed a formal technique to compare different materials based on the function required of the material. The principles that he introduced combined the structure of scientific method with detailed analysis of the functions to be performed. This process was refined and developed by Miles into the value analysis technique.

While value analysis was originally applied to the re-design of existing products, the scope of application and the range of techniques available have, over time, evolved into the value management systems used today to optimize the value of projects, products and business processes across a broad range of commercial and public sector activities. Much of the thrust of value management in recent years has been focused on the creation of competitive advantage. Although significantly developed over the years, the principles developed by Miles still lie at the heart of value management.

Definitions

Value

Value is a complex concept. It is a measure of worth – a relative measure of the usefulness of something in relation to the cost paid for it. It can be expressed as a function

$$\text{Value} = \frac{\text{Function}}{\text{Whole life cost}} \quad \text{or} \quad \frac{\text{What you get (or want)}}{\text{What you pay}}$$

However, value is like beauty insofar as it depends upon who is judging it. Value is very much about perspective. This is what makes the concept of value complex in a business context. Businesses and organizations tend to be multi-faceted comprising a wide variety of different groups each with their own interests, priorities and perceptions of what is required. In this context the likelihood of ill-conceived projects, waste and non-congruent business processes, is high. An operations manager, for example, may have a completely different view of the value of a new production process to that of a marketing manager or a human resource manager.

However, the individual requirements, perspectives and values of all stakeholders, aligned with the strategic objectives of a business or organization, are important ingredients in setting objectives for business processes, capital investments and continuous improvement. The procedures of value management bring all key stakeholders together and try to define congruent objectives against which value can be measured and improved.

The optimization of value is addressed through the process of function analysis. The process is aimed at identifying, and synthesizing through consensus, the objective criteria required from a business process, an asset, a product or a project.

Causes of poor value and waste in business include

- Politics
- Stakeholder rivalry (lack of congruence)
- Lack of information
- Inappropriate/insufficient briefing
- The prevalence of habits, attitudes and customs
- The adoption of temporary solutions
- Management complacency.

The objective, team approach of value management helps overcome these problems, improve value, reduce waste and provide competitive advantage.

Value management

Value management is a structured, multi disciplinary group decision-making process that provides a planned approach to the enhancement of the value of a project, process or product, in a manner that is consistent with broader strategic or business goals.

Value management is the generic term for the full range of value techniques and procedures. The techniques can be classified into value planning, value engineering and value analysis.

The value process works top-down, starting with needs and strategic goals and focusing on root causes, not symptoms. It is a flexible team-based activity often conducted by an independent facilitator. It is driven by consensus and a workshop approach is usual. The workshops are generally short in duration (usually one to two days), intense and highly structured. The process does not impose solutions; the outcomes and decisions are those of the workshop team, resulting in total ownership by the team.

Value planning

Value planning establishes an early consensus between stakeholders about the need for a capital project, its scope, the deliverables, the key functions and risks, all in the context of overall business or organizational strategy. It explores opportunities for innovation and identifies the most cost effective means of implementation, consistent with desired time and quality requirements. Value planning considers the whole and its relationship to business need rather than the individual components.

Value engineering

Value engineering is a procedure aimed at providing an organized, systematic process of technical appraisal of a developing project, product or process to eliminate unnecessary costs and add value while maintaining or enhancing quality, scope and performance.

The objectives of value engineering studies vary but commonly include

- achieving agreement on a project's objectives and deliverables;
- selecting the best design from the available options;
- optimizing the value of a chosen option;
- maximizing the efficiency of a chosen design and its delivery.

Value analysis

Value analysis is the application of the procedures to an existing project, product or process. It is a source of benchmarking and can act as a structured basis for achieving continuous improvement.

The value process

The value process generally involves putting a multi-disciplinary team through a structured and logical sequence of tasks with the overall objective of ensuring that the optimum value to the business or organization and its stakeholders is represented by the product, project or process under development. The sequence of logical steps has remained broadly unchanged over the years since its initial development by Larry Miles. It remains applicable today. The steps are as follows:

Step 1: Initiation

The initiation phase answers the questions 'What?' 'Who?' 'Where?' and 'When?' It is analytical and involves the key stakeholders in the study exploring what is being studied and why, 'What' they hope to get out of the study, 'Who' should be involved, 'Where' it should take place and 'When'.

Step 2: Function Analysis

The formal analysis of function lies at the heart of the method and answers the question 'Why?' Why certain components are included and what contribution they make to the end purpose of the object of the study; the mode of thinking during this stage is still analytical. It is important at this stage to achieve consensus on functional requirements and to test that the requirements can be related to overall corporate or organizational strategy.

Step 3: Speculation

The speculation stage poses the question 'How?' How else can the function(s) be achieved? It is used to generate alternative ideas and provides the innovative springboard of the method; here, the team will employ creative thinking without evaluation.

Step 4: Evaluation

Generated ideas are evaluated to answer the question, 'Is this better than what we had before?' During this phase the team's thinking is judgemental. It will include consideration of opportunity and risk.

Step 5: Development

In step 5 the team answers the question, 'Will it work?' Once more the team is back in analytical mode and must validate the proposals that have been selected and appear to be better alternatives to the original.

Step 6: Implementation

It is in the implementation phase that the developed and validated proposals will be implemented.

Step 7: Audit

Audit is the final step and answers the question, 'Did it work as well as we expected?' Again, this is analytical and provides feedback for continuous improvement.

The application of value management

It is possible to apply the techniques in an *ad hoc* way, having workshops as and when deemed appropriate and necessary by management. Many organizations adopt this approach and, indeed the evidence suggests that generally the approach is positive and better projects, processes and products result.

An alternative approach involves the creation of an organizational value culture to continuously improve all aspects of business from corporate strategy through the management of operations. The techniques and tools of value management can be used to underpin and make possible this culture. The general move seems to be towards the adoption of this value culture; indeed similarities can be drawn between this approach and the lean production processes used successfully in Japan over the last few decades. However it should be remembered that culture develops and evolves over many years and for real benefit the culture must be viewed and perceived as a permanent change necessary for the future success of the organization and not simply as a passing management fad, that will be replaced by the next management fad in a few years time.

With regard to involvement in the processes of value management the following three points are key to success.

- Firstly, it is essential for senior management to demonstrate and express their support and commitment to the use of value techniques – without this support it is often impossible to get the appropriate level of commitment and buy-in from elsewhere in the organization to ensure success.

- Secondly, the facilitation and running of value programmes and workshops can either be undertaken from within the organization or by the use of external facilitators – the decision is really down to available resources and the frequency of value studies within the organization. It is however essential for those facilitating to have the necessary skills and to demonstrate impartiality to the stakeholders involved.
- Thirdly, it is essential that all key stakeholders in the project, product or process under examination are involved.

With regard to optimum timing for studies, it must be recognized that for developmental projects the opportunity for a change and hence for a value improvement is considerably greater during the early stages of development than at the later stages. Consequently, as a priority, value techniques should be used to validate project need, project brief and initial design solutions.

However, where the techniques are used to support a value/ continuous improvement culture the timing issues and strategy will be different.

The benefits of value management

The use of value management techniques and the establishment of a value culture encompassing the techniques can deliver the following benefits

- The need for new investment is always verified and project goals are clearly defined.
- Objectives and decisions are openly discussed and explicitly stated.
- Evaluation frameworks are structured, rational and rigorous.
- Business decisions are supported by data and made on the basis of defined performance criteria aligned with corporate strategy.
- Accountability is increased.
- Alternative solutions are always sought and considered.
- Business decisions can be made with greater confidence.
- There is enormous potential for increasing value for money.
- Communication, understanding and teamwork can be improved and disseminated throughout the organization.

- Participation by all key stakeholders increases the likelihood of satisfaction with the end product.
- Opportunities for long-term profitability and continuous improvement are enhanced.

Some value management case studies

The following case studies demonstrate how value management may be applied.

Case 1 – Implementing a value management programme

An international manufacturer of a wide range of consumer products spends over $500 million/annum on capital projects to improve its manufacturing capabilities. Business analysts noticed that the productivity of the assets was falling behind that of its competitors. They sought to implement a value management policy to redress this balance and improve asset productivity by a target figure of 20% over a period of two years. An individual within the central engineering department was made responsible for delivering this value management programme, which included undertaking a value engineering study on every capital project exceeding $5 m. During an 18-month period over thirty studies were carried out, achieving savings of between 5 and 30%. The success of the programme was evidenced by the fact that it made a significant impact on the company's market capitalization.

Case 2 – Optimizing the benefits of joint venture projects

Two major, non-competing, financial service organizations joined forces to develop a combined operations centre. The rationale for the joint venture project was that a combined centre would lower operating costs and hence create competitive advantage for both organizations. A joint brief had been developed by the organizations and a sketch design solution evolved in response to this brief. The joint venture commissioned a value management study involving all the major stakeholders to optimize the value of the brief and design solutions to the benefit of the joint businesses. A one day workshop study was conducted which identified that, far from reducing costs, the brief required a facility that met all the conflicting corporate, process and security requirements of each organization – duplication and confusion abounded. The workshop identified brief and design changes resulting in a re-design of the project representing a 30% reduction in capital cost, a significant reduction in operating costs and the realization of competitive advantage for both organizations. The study succeeded because the major stakeholders for the businesses and the project were present at the workshop and consequently owned the solutions developed during the workshop.

Case 3 – Re-engineering the procurement process

An oil and gas company desired to review the process by which it conducted its business of design, procurement and contracting of off shore production facilities. Three areas of focus for reducing cost were identified as follows:

- Functional performance specifications
- Reduction in procurement paperwork and
- Simplification of procurement related processes.

A value management study was instigated involving members from all key disciplines and including representatives from most levels within management. The company prides itself on teamwork and this contributed enormously to the effectiveness of the study. The study was a success and many proposals were implemented quickly while others depended upon other changes taking place before their potential could be fully exploited. In many cases improvement in performance exceeded 50%, dramatically reducing man-hour requirements, delivery times and documentation but, at the same time, enhancing product quality. With the help of the value techniques adopted the company enhanced its competitive edge.

Case 4 – Validating the project brief

This organization made it a condition of its headquarter relocation and consolidation project that a value management programme be implemented to achieve optimum value for money and to ensure that the project comprehensively addressed functional and performance requirements. The project involved consolidating numerous business sectors on over fifty sites to less than ten carefully chosen new and existing locations. There were strict budgetary limits and spend profiles for the project. A series of twenty value workshops were conducted, using a single functional model for consistency, to validate that the outline briefs and initial design concepts accurately reflected the organization's requirements. The results of the workshops contributed significantly to the re-planning of several of the sites and provided the basis for much improved definition of quality and functionality.

Case 5 – Building the team

A major organization, with a high public profile dealing with sensitive issues, wanted to ensure that its buildings were developed to the highest standards and fully utilized the limited resources available. The organization had formally incorporated value management into its procurement procedures.

One particular study was undertaken after the appointment of the building contractor and scheme design but before the detailed design of a design, develop and construct contract. All parties were represented at a value workshop, including the organization's design team who retained a 'watching brief' over the design development by the construction team. It soon became apparent that there was a difference in design philosophy between the two teams; both were very experienced and put forward excellent and workable proposals – the problem was that they were different. The atmosphere of the workshop proved to be the ideal forum to explore both options in depth and understand the others viewpoint. By the end of the workshop, the best parts of both schemes were adopted which far surpassed either original scheme. The study started with two teams and two philosophies; it finished with one team and one far better scheme.

Case 6 – Achieving NPC savings

Many value engineering studies are concerned with obtaining better value for money often involving reducing costs, preferably measured on a life cycle basis over the design life of an asset. This case involved the examination of a proposed chemical packaging facility. The primary aim of the project was to replace an existing, but worn out facility, with a modern one to comply with ever more stringent quality control legislation and to improve competitiveness. To save capital costs, it was proposed to reuse and extend the existing building to fill the available site but replace all the process plant within it with state of the art equipment. To save running costs, it was proposed to automate the plant as much as possible, particularly all handling equipment.

The result of a two-day study workshop was a saving of 50% in the NPC of the plant. The principal savings arose from just one proposal, namely to design the optimum process-flow path and then provide a new building to house it. This enabled considerable simplification of the entire facility reducing both capital and running costs.

The benefits in summary

In summary some of the key opportunities of value management are

- To help optimize the balance between corporate strategy, market threats and opportunities and investment decisions and to help prioritize capital investment.
- To validate and optimize the value of capital project briefs and design solutions and to help evolve the design and construction of capital projects to meet functional requirements at lowest whole life cost.
- To help establish the optimum procurement and funding strategy for capital investment.
- To define, design and continuously improve business processes and the physical infrastructure supporting these processes to deliver world-class solutions at the lowest whole life cost.
- To establish the strategies and relative economies associated with safeguarding for future change.
- To strike the optimum balance between capital and operating costs.
- To benchmark the performance of existing facilities and to identify potential value improvements that can be applied to new facilities.
- To realize and maximize the value associated with business partnerships and strategic alliances.
- To support continuous improvement in operating processes through value analysis.

A function led value management approach can be used to identify and optimize investment for all types of organizations and act as a basis for continuous improvement through improving revenues while reducing operating costs.

LIVERPOOL JOHN MOORES UNIVERSITY
LEARNING SERVICES

19

Insolvency

Introduction

Unfortunately, insolvency is a common occurrence within the UK construction industry. Its causes are various. Certainly, the ease with which contracting businesses can be created, due in part to the minimal amount of capital investment required to start a construction firm, often gives rise to some very fragile arrangements. The capital goods nature of construction industry products, together with the large proportion of industry output flowing from often widely fluctuating and unpredictable levels of public expenditure, has created a 'boom and bust' construction industry cycle. This is often cited, particularly at times of recession, as a cause of mass insolvency.

Competitive tendering practices used throughout the industry have also been cited to explain the high incidence of contractor insolvency. The quality of management expertise within the industry is generally considered poor in comparison with most other industries. There are, certainly, many cases of contractor insolvency which have resulted from the mismanagement of a business as opposed to unfavourable external conditions. Whatever the causes of insolvency, however, the situation is unlikely to change in the short term. Surveyors must, therefore, be equipped to deal with the incidence of contractor and sub-contractor insolvency.

The damage resulting from insolvency can be widespread. The cost, quality and duration of a construction project can all be detrimentally affected as a consequence. Additionally, looking down the contractual hierarchy, the solvency of sub-contractors and suppliers can also be damaged following the insolvency of a contractor; thus one insolvency can affect many projects. The failure of one large British contracting organization during the recession of the 1980s resulted in a long chain of insolvencies of associated sub-contractors and suppliers. The insolvency of a developer can create financial difficulties for the contractors and, consequently, the sub-contractors and suppliers involved with his developments. Again, these problems can manifest themselves on other projects for other clients.

It is important for surveyors to ensure, as far as they can, that the contractors and sub-contractors they appoint are financially stable and are likely to remain so for the duration of their contracts. Obviously, this can never be guaranteed, but it is prudent to make formal and informal checks prior to contract or, preferably, tender. In the unfortunate event of contractor insolvency during the course of a project, surveyors must advise and act to protect the interests of their clients. This will mean ensuring that projects are completed, where appropriate, with minimum loss of value to their clients in terms of increased time, additional cost and reduced quality (see section on *Methods of Contractor Selection*, p. 69).

This chapter provides a brief introduction to the law of insolvency. It examines the provisions in the JCT Standard Building Contract insofar as they apply to the insolvency of either the contractor or the Employer. It considers the action and procedures to be adopted leading up to, and following, contractor insolvency. Finally, it discusses the situation of insolvency and domestic sub-contractors.

The law

The main statute covering insolvency law is the Insolvency Acts 1986 and 2000 (the Act). The 1986 legislation was introduced to implement some of the recommendations of the Cork Report. The Act has improved insolvency practice by controlling those persons entitled to act as insolvency practitioners and by introducing two new procedures, the administration order and the voluntary arrangement, which, as Newman[1] notes, 'aim to preserve a business in some form or other. This is a major departure from the emphasis of traditional insolvency procedures on extinguishing businesses'. There are, of course, many other statutory instruments and the like, together with much case law, pertinent to insolvency. It is not within the scope of this chapter or, indeed, a book of this nature to identify and explain the law in detail. Nor is it essential for the majority of surveyors to have this level of detailed knowledge. However, this section will provide a general introduction to the legal meaning of insolvency which, in itself, is often a source of confusion.

As the majority of contracting organizations are companies of one form or another, this chapter deals with the law relating to corporate insolvency. It should be noted that the law and procedures are different for individuals and partnerships.

Insolvency is broadly concerned with the inability to pay debts. There are various situations which can arise under the Act.

(i) Liquidation

This refers to the winding-up of a company. Trading usually ceases upon liquidation; the assets of the company are collected and used to offset liabilities. There are two categories of liquidation – *voluntary liquidation* and *compulsory liquidation.*

Voluntary liquidation can be either members' or creditors' voluntary liquidation.

Compulsory liquidation is the result of a court order for a company to be wound up. This is usually because the company is unable to pay its debts. There are two tests that can be used to determine if a company is unable to pay its debts. First, the going concern or cash flow test, whereby a company is judged to be insolvent if it cannot pay debts as they become due. Secondly, the balance sheet test, which is a more long-term test. The balance sheet test examines the value of a company's assets in relation to the amount of its liabilities. If assets are less than liabilities, then the company will be deemed insolvent. The nature of construction – in particular the difficulties of accounting for work in progress – can make the application of this second test difficult.

A liquidator will be appointed to take control of the winding-up of the business. For compulsory liquidation this will, in most cases, be the Official Receiver. The proceeds of liquidation are distributed in accordance with the following hierarchy:

(a) fixed charge holders;
(b) liquidators' fees and expenses;
(c) preferential creditors – pension schemes; employees' pay; etc. (with the introduction of the Enterprise Act the Crown has

given up its previous preferential creditor status in relation to PAYE, VAT and National Insurance contributions);
(d) floating charge holders;
(e) unsecured creditors;
(f) shareholders.

Once a winding-up petition has been presented by creditors, they can apply to a court for a provisional liquidator to be appointed.

(ii) Administrative receivership

Prior to the introduction of the Enterprise Act it was not uncommon to have an administrative receiver appointed to a construction related company that was in financial difficulties. An administrative receiver could be appointed by a debenture holder (usually a bank) in accordance with a provision in the debenture agreement regarding a specified situation occurring. A debenture is a security given to a lender against borrowings. The duty of the administrative receiver is to recover the borrowings owed to the debenture holder by realizing those assets which are the subject of the debenture. In some circumstances, it may be advantageous to the debenture holder for the administrative receiver to keep a company operational if more of a debt may thus be realized. However, the Enterprise Act has now severely limited the rights of debenture holders and others from appointing administrative receivers by introducing an amendment to the Insolvency Act 1986, i.e., 'The holder of a qualifying floating charge in respect of a company's property may not appoint an administrative receiver of the company'.[2] There are a few limited exceptions to this rule that still allow the appointment of an administrative receiver, for example a project company set up under a Public–Private Partnership agreement which includes step-in rights. In the majority of instances where a debenture holder wants to take action against a company in financial difficulties it will now have to consider the use of an administration order.

(iii) Administration order

The idea of administration was introduced into the UK by the Insolvency Act 1986, and was based to some extent upon processes which had been used with some success in the USA. The main aims of administration are to create a culture of company rescue or alternatively to try and maximize the realization of the company assets and allow a fairer distribution amongst the many classes of creditors.

There are three ways by which an administrator may be appointed to a company:

Appointment by the company or directors whereby they file, with the court, a notice of intention to appoint an administrator. At the same time as filing the notice with the court the directors must send a copy to every holder of a qualifying floating charge, seeking their written consent to the appointment before it may be executed.

Court appointment of administrator. If a creditor of a company wants to initiate the appointment of an administrator he must make an application to the court and obtain a court order. A court will issue an administration order if it is satisfied that a company is either unable, or likely to become unable, to pay its debts, but only if a benefit is likely to arise as a consequence of the order (for example, the survival of the company or a better financial position than would be achieved with liquidation). The creditor must also submit a copy of the application to any other organization that may be able to appoint an administrator, e.g., companies with qualifying floating charges.

Appointment of an administrator by the holder of a qualifying floating charge. A holder may file a formal notice of intention to appoint an administrator at the court. A copy of this notice must be sent to any organizations holding prior qualifying floating charges and a request for their consent to the appointment. If there is no dissent a notice of appointment must be filed with the court within five business days of the issue of the first notice.

Dissent to appointment of an administrator

In either of the first two instances above a holder of a qualifying floating charge has a right of objection to the appointment of a proposed administrator and has the right to appoint an administrator of their choosing. Similarly in the third instance a holder of a prior qualifying floating charge can object to the administrator proposed and has the right to select their own administrator if they wish.

An administration order freezes the affairs of a company, thus allowing breathing space for a corporate rescue to be investigated.

Once an order has been made, a company will be managed by an administrator who will discharge the duties of the directors. It is the administrator's duty to manage the company in the interest of the creditors. Administrators have wide-ranging powers, and are generally permitted to do anything necessary to manage the business. Within eight weeks of the date the company entered into administration, the administrator must submit proposals to creditors and company members with regard to achieving the purpose of the administration order, and he must hold an initial creditors' meeting within ten weeks from the date of administration. A time limit of one year is placed on the administration process although this may be extended by the creditors and/or by the court.

(iv) Voluntary arrangements

These enable the directors to make a proposal to their company and its creditors with regard to the payment of debts. The creditors are invited to a meeting at which a decision is made with regard to the acceptability of the proposed arrangements.

Client's creditor status

Where a contractor is put into liquidation or administration the client's position will normally be that of an unsecured creditor. In many situations the end result for the client is that once all the preferential creditors have been paid there will be little or no funds available to pay any outstanding claims the client may have against the contractor. The Enterprise Act has slightly improved the position of unsecured creditors by the removal of a substantial part of the Crown's preferential status. This means that more funds will be available to creditors further down the order of distribution of funds. The Enterprise Act also requires that where funds permit, a liquidator or administrator must reserve a portion of the realized assets for the benefit of unsecured creditors.

Provisions in the JCT Standard Building Contract

Section 8 of JCT 05 contains the provisions dealing with either the insolvency of the contractor or the Employer. At the commencement of this section the JCT provides a definition of insolvency for use within these contract conditions. For the purpose of JCT 05 a party is considered to be insolvent where the following apply:

1 they enter into an arrangement, etc., in satisfaction of their debts (i.e., see voluntary arrangements above);
2 they pass a resolution for their company to be wound up, without making a declaration that the company is solvent. For example, where a solvent company wishes to cease trading and be wound up they would be in a position to make a formal declaration that sufficient assets exist to pay off any liabilities therefore creditors would not be too concerned about such a resolution;
3 a winding up order or bankruptcy order made against them;
4 an administrator or administrative receiver is appointed;
5 they are subject to similar insolvency proceedings outside the UK;
6 in the case of a partnership, all partners are the subject to an individual arrangement or any of the above events.

Contractor insolvency – (clause 8.5)

If any of the above insolvency events apply to the contractor he is obliged to inform the Employer through a written notice. Where the contractor is insolvent the Employer may terminate the contractor's employment. It is important to understand that under this procedure it is only the contractor's employment that is terminated, the contract conditions are still operational and will be used to resolve the problems caused by the termination. If the Employer decides to terminate the contractor's employment he must send the contractor a written notice of termination. The notice must be delivered by hand, special delivery or recorded delivery, in this instance an electronic communication would not be acceptable. The termination will become effective from the moment the contractor receives the notice. The contractor (through the insolvency practitioner) and the Employer can, however, agree to reinstate the contractor's employment using either the existing contract conditions or alternative terms that the parties may agree.

As from the date of the contractor's insolvency, and whether or not the Employer has issued a termination notice, the parties obligations under the contract conditions are amended. The Employer is no longer obliged to make further payments or release retention to the contractor, however if an interim certificate has been issued a few days before the date of the contractor's insolvency an Employer would be advised to issue a withholding notice for the amount due in that certificate and not rely on clause 8.5.3.1 (*Melville Dundas Ltd v George Wimpey UK Ltd, 2005*). The contractor's obligation to carry out and complete the works and design work for any contractor's design portion is suspended. The Employer is also empowered to take reasonable measures to protect the Works and site materials during this period, the cost of such measures being deductible from sums due to the contractor.

Termination of contractor's employment – (clause 8.7)

Having been notified of the contractor's insolvency the Employer may wish in such circumstances to appoint a new contractor to complete the Works; in this event, the right to terminate the contractor's employment is exercised. This decision may depend upon the specific requirements of the Employer, the stage of the project and the likelihood of suitable alternatives being offered by the administrator. However, if the Employer reserves this right of termination, there may be more scope for the administrator to sell the contractor's business, including contracts in progress, to another contractor, which might be an advantageous course of action for an Employer. Care should be taken, in exercising the right to terminate, to ensure that an Employer's losses are not increased above what may be offered by the administrator.

The consequences of termination by the Employer

Clause 8.7 details the contractual consequences of termination by the Employer. The Employer is granted the right to employ other persons to complete the Works (and design where applicable) and to make good defects. Use of materials, plant, etc., on site is permitted with the proviso that, where they are not owned by the contractor, the consent of the owner is obtained. The requirement for the contractor to assign the benefits of supply and subcontract agreements to the Employer upon termination may not be possible in respect of termination flowing from contractor insolvency (*see* JCT 05, footnote 58). The contractor must remove any temporary buildings, plant, tools, materials, etc., which belong to him and make arrangements to have items not owned by him removed if so requested by the architect.

It is common for materials to be supplied to a contractor on the condition that ownership of the goods does not pass to the contractor until the goods have been paid for. Such a retention of title claim (or Romalpa clause) will fail if the materials have been fixed in place. Further protection from retention of title claims is provided, by section 25 of the Sale of Goods Act, when the Employer has paid for them in interim payments (*Archivent Sales Developments Ltd v Strathclyde Regional Council (1984) 27 BLR 98*). However, the situation is not so clear in respect of sub-contractors' unfixed materials as these materials are not covered by the Sales of Goods Act.

The financial settlement, assuming the Employer completes the work, is calculated in accordance with the provisions in clause 8.7.5. The Employer is not required to make further payment to the contractor until the Works are complete, defects are made good, and the account has been prepared. The account should detail:

(i) the total expenses to the Employer following termination (including any direct loss or damage caused as a result of the termination);
(ii) payments made to the contractor;
(iii) the amount that would have been payable had the Works been completed under the contract (i.e., a notional final account).

If the sum of (i) and (ii) is greater than (iii), the difference is a debt payable by the contractor to the Employer. If less, the difference is a debt owed to the contractor by the Employer. This principle is illustrated in Table 19.1.

Clause 8.8 deals with financial settlement where the Employer elects not to complete the Works.

Employer insolvency

In the event of Employer insolvency, a contractor is empowered, under clause 8.10.1, to terminate his employment under the contract. The consequences of termination are set out in clause 8.12; they require:

(a) the contractor to remove (and ensure that sub-contractors remove) temporary buildings, plant, materials, etc., from the site;

Table 19.1 Example of claim against contractor in liquidation

Statement of claim by City Developments Developments Ltd v Eagle Construction (in liquidation). Re. Park Marina project		£
Notional final account for Eagle Construction (iii)		3,784,960.00
Amounts paid to Eagle as Interim Certificate no. 7 (ii)		1,492,380.00
Amounts paid to LKA Contractors for completion (ii)		2,576,440.00
Direct loss and damage (i)		
Temporary security	6,700.00	
Temporary insurance	1,450.00	
Additional fees incurred through termination	4,370.00	
Additional tendering costs	3,550.00	
Liquidated damages directly attributable to Eagle Construction	20,000.00	
Credit: sale of plant and temporary buildings owned by Eagle Construction	(2,350.00)	33,720.00
Sub-total		4,102,540.00
Amount due to City Developments from Eagle Construction		317,580.00

(b) where there is a contractor design portion, the contractor is to provide the Employer with 2 copies of the design documentation identified in clause 2.40, as far as it has been prepared by the contractor at the date of termination;

(c) the contractor to prepare an account detailing:
   (i) the total value of work completed and any other sums due under these conditions;
   (ii) sums in respect of direct loss and/or expense payable under clauses 3.24 and 4.23;
   (iii) the cost of removing temporary buildings, plant, materials, etc., from the site;
   (iv) direct loss and damage caused by the termination;
   (v) the cost of materials and goods ordered and either paid for by the contractor or which the contractor is legally bound to pay for; ownership of these materials will pass to the Employer on payment by the Employer;

(d) the difference between the account and the amounts paid to the contractor is to be paid by the Employer within 28 days of presentation of account (no Retention is to be deducted). In practice, termination following insolvency may make this payment unlikely. In such circumstances the contractor will become an unsecured creditor of the Employer.

An Employer's interest in Retention is as a trustee for the contractor. JCT 05 clause 4.18.3 requires the Employer (local authorities are not bound by this condition) to place retention monies in a separate bank account (designated to identify the amount as Retention held by the Employer on trust) if so requested by the contractor. In the event of Employer insolvency, the separate bank account will ensure the trust status of Retention and should therefore protect a contractor's right to the Retention. In the absence of a separate account, case law suggests that a contractor may lose this right. It is, therefore, strongly recommended that, as a matter of course for all contracts, contractors should request that Retention be held in a separate identified bank account by the Employer.

Avoidance, precautions and indications

Avoiding the incidence of contractor insolvency is obviously to be preferred. Where selective tendering is the basis for a contract, the pre-selection process should endeavour to check the financial status of those contractors under consideration. With open tendering, the checks must be made during the tender appraisal stage. There are no hard and fast rules regarding the checks to be made. *The Code of Estimating Practice*[3] gives guidance on pre-selection good practice. The following are common sources of information which can help to assess a contractor's financial standing:

(i) information that members of the design team may have regarding individual contractors;
(ii) information from financial checking agencies;
(iii) published accounts;
(iv) bank references;
(v) other references.

Precautions can be instigated to soften the blow of contractor insolvency. These may include the use of performance bonds and guarantees (as discussed in Chapter 17) and contractual Retention funds (most standard forms of construction contract prescribe retention of a proportion of interim payments pending satisfactory completion).

The signs of impending insolvency often become apparent some time in advance. If the signs are spotted then, at best the situation may be avoided, or, at least the impact of the insolvency may be reduced by taking pre-emptive action. In the run up to, and during the course of, a contract, surveyors should be aware of any signs that may signal trouble. These may include:

(i) industry rumours;
(ii) slowing down of progress on site;
(iii) complaints of non-payment from sub-contractors and suppliers;
(iv) reduction in personnel on site;
(v) reduction in amount of materials on site;
(vi) over-ambitious requests from a contractor for payment.

Surveyors must be careful in acting upon these signs. Impending insolvency can be accelerated, or sometimes caused, by the spreading of rumours suggesting financial instability. Once news gets around, creditors soon start calling in debts, and thus a contractor's difficulties can be intensified. So, if there are concerns, they should be kept within the design and client team until such time as the situation becomes official.

Action in the event of contractor insolvency

Action on suspicion of contractor insolvency

If the design team suspects that a contractor is suffering financial difficulties, there are some measures that should be considered to safeguard the Employer's interests should these suspected difficulties result in insolvency. Interim valuations should be carefully prepared to ensure that work is not overvalued. Materials on site should be carefully checked to ensure that they are in accordance with the contract, properly stored, intended for incorporation into the Works and that it is not unreasonable for them to be on site in relation to the programme. Suspicion does not give the right to undervalue work, but it would be prudent for surveyors to pay more attention to ensuring the accuracy of their interim valuations in such circumstances.

Immediate action upon contractor insolvency

It is important to act quickly following the announcement of contractor insolvency. News of insolvency can travel very quickly, and creditors, in an attempt to mitigate their losses, may try to recover debts by removing materials and equipment from sites. The following measures should be considered by the design team:

1. Advising clients of contractor insolvency, the contractual position, recommended action and liability.
2. Securing the site and considering employing a security firm to provide 24-hour security, changing the locks, carefully securing all valuable equipment, goods and materials.
3. Preparing a detailed valuation of completed work and an inventory of materials and equipment on site.
4. Stopping the processing of any payments to the contractor.
5. Contacting key sub-contractors and suppliers and commencing discussions concerning continuation contracts.
6. Checking available bonds and guarantees and, where appropriate, following the prescribed procedures (e.g., informing bondsmen).
7. Contacting the administrator or liquidator and eliciting their views with regard to completion of the project.
8. Keeping a record of all time spent and costs incurred in dealing with and advising on the insolvency. It is normal for additional fees to be chargeable in this respect.

Completing the works

There may be a number of options open to an Employer with regard to completing a project following the insolvency of a contractor. Surveyors should advise their clients of the alternative strategies available, together with the likely implications of each of these strategies, thus enabling their clients to make informed decisions. The alternatives available for completion may include:

Continuing with the original contractor

This would also include reinstatement of the contractor's employment where termination had occurred. This is usually only practicable where a project is near completion and the contractor is able to continue.

Assignment or novation of the contract

If the administrator is able to sell off some of the insolvent contracting organization's contracts, then assignment or novation of the contract with the purchasing contractor may be an attractive route for completing a project. *Assignment* is the legal term used to describe the transfer to a third party of a contractual right or obligation. The legal position is complex, and provision for assignment under JCT contracts is confusing. Novation is often preferred by both administrators and Employers. *Novation* involves forming a new agreement with a new contractor for completing the work. The new contractor will 'step in' to replace the insolvent contractor and take over its obligations to complete the Works.

Appointing a new contractor to complete the project

In some situations it may be prudent to appoint a new contractor. This choice may depend upon the stage of progress on a particular project. If a contract has just commenced prior to insolvency, it may be possible to approach the second tenderer. On the other hand, if a project is near completion, and in the absence of an indication from the administrator that assignment or novation may be possible, appointing a new contractor may be the only course of action that will ensure the completion of the project.

When faced with making, or advising on, such a decision, it is important to:

(a) consider all the factors that affect the decision and select the strategy that offers most benefits;
(b) act quickly to ensure that losses are minimized;
(c) keep all options open until a formal agreement for completion has been made.

The basis of continuation contracts

Where a new contractor (or a series of contractors) is to be appointed to complete a project, it will be necessary to consider the basis of the contract and the documentation to be used. Generally, the options available will be the same as those for all projects, although the decision will be affected by:

(i) the time available to prepare documentation and agree terms;
(ii) the scope and amount of work involved;
(iii) the progress made by the original contractor;
(iv) the documentation and contractual basis of the original contract;
(v) the need to obtain competitive tenders.

Where a project commenced shortly before contractor insolvency, it may be possible to use the original tender and contract documentation with an addendum to take account of the work completed. In this situation, it may also be possible to negotiate a contract with one of the original tenderers for the project, preferably on the basis of their original tender. It may be appropriate in other situations to let the work on the basis of a prime cost contract, particularly where the scope of work is small and time is critical. If the only work required is to make good defects, the appointment of jobbing contractors as and when required may prove effective. It is not necessary to let completion contracts on the same basis as the original contract. The specific circumstances pertinent at the time the completion contract is being considered should indicate the most appropriate contractual route.

Domestic sub-contractors

The effect of contractor insolvency

If the employment of the contractor is terminated, then the employment of any domestic sub-contractor is automatically terminated (*see* clause 3.9.1). On some projects, the work of a domestic sub-contractor may be fundamental to the satisfactory completion of the project, e.g., they may have been responsible for designing an important aspect of the building and it would be extremely difficult for another contractor to move in and take over the design and associated work. If the work, or the status of the work, of a sub-contractor is such that it would be most appropriate (for the completion of the project) for the sub-contractor to be involved in the continuation contract, then this involvement should be explored with contractors tendering for the completion of the works. There may, however, be some problems in practice. The sub-contractor may be owed money in respect of work undertaken and may, in the absence of payment, not be prepared to continue. There is no provision in the contract conditions for domestic sub-contractors to receive a direct payment from the Employer. If the Employer were to make a direct payment to a domestic sub-contractor for the sums not paid by the main contractor it is almost certain that the Employer would not be able to reclaim the money as a cost against the insolvent contractor. An administrator or liquidator would claim that such payments were contrary to the insolvency laws.

References

1. NEWMAN, P., *Insolvency Explained* (London: RIBA Publications, 1992).
2. Enterprise Act 2002, Chapter 40, section 250, sub-section (1) (London: HMSO).
3. Chartered Institute of Building, *Code of Estimating Practice*, 5th edition, Englemere, 1983.

Bibliography

1. NEWMAN, P., *Insolvency Explained* (London: RIBA Publications, 1992).
2. STEWART, A. and BILLINGHAM, E., Hands off! Receiver, *Building*, 13 August 1993, pp. 16–17.
3. PHIPPS, M., Plugging those insolvency gaps, *Building*, 9 October 1992, pp. 30–31.

20

Dispute resolution

Introduction

Disputes are a common occurrence within the construction industry. One has only to examine reported legal cases to get a feel for the number of disputes. Of course, this is only the tip of the iceberg. Many disputes are resolved before they reach this stage, either on 'the courtroom steps' or through arbitration or by some other method of resolution.

Relationships within construction, between clients, contractors, sub-contractors and suppliers, are often adversarial. The risks associated with construction projects can be high, the process is complex and obligations are often onerous. The way demand is put on the industry through competitive tendering procedures can often increase adversity. The need for contractors to 'win' contracts can often mean that a project gets off to a bad start, with the contractor battling to ensure cost recovery.

Disputes are damaging both to the industry and to the clients of the industry. The costs associated with resolving disputes are often high, and the time involved can be very long. The overall impact of disputes on individual projects can jeopardize the objectives of all involved in them. A preventative approach would obviously be the best situation, that is, removing the causes of disputes altogether. However, such an environment could only be created by radically changing the processes, attitudes and structures that lead to dispute. It is unlikely in the short term (if at all) that such changes will be made. There will, therefore, continue to be a need for disputes to be resolved.

This chapter examines the various methods and approaches to resolving disputes in connection with construction projects. It covers litigation, arbitration, alternative dispute resolution and adjudication.

Litigation

Traditionally, disputes that could not be resolved directly by the parties involved in the dispute were either dealt with by the courts or referred to arbitration. *Litigation* involves one party issuing a writ against another. The parties are required to prepare a case in support of their argument, and this is heard in court. The mechanics of litigation are governed by statute. Decisions are made on the basis of the evidence presented, together with the application of relevant legislation and precedent. The process is often long and expensive, and can be damaging to business relationships. The role of the surveyor in litigation will normally be as a witness of fact or as an expert witness. Expert witnesses may be called by a party to a dispute to give expert evidence of opinion to support an argument.

Litigation is often an ineffective way of deciding disputes in the context of business relationships. Court proceedings are more concerned with deciding a winner and a loser to the dispute rather than establishing a compromise solution, which reflects the business needs of all concerned. Litigation is adversarial and will tend to increase the conflict between parties to a dispute. Such an adversarial approach can damage a business relationship; in the context of construction, this can cause untold damage to the successful completion of an ongoing project.

Arbitration

Arbitration is the process whereby the parties refer any disputes arising in the course of a contract to an agreed third party to be resolved. Where reference to arbitration is in writing, the procedure is governed by the various Arbitration Acts. The Acts require arbitration to be carried out in a judicial manner, and at the conclusion of a hearing the Arbitrator must provide reasons for any award made. The decision of the Arbitrator in relation to findings of fact is conclusive; there is no route for appeal. However, appeals can be made to the courts on points of law, but only with the approval of the court hearing the appeal. The award of an Arbitrator is legally binding.

The advantages of using arbitration as opposed to litigation for resolving disputes include

1 *Cost* – arbitration is generally much cheaper. However, the costs of arbitration are significant, and it is widely felt that the cost benefit has been eroded over time.
2 *Speed* – arbitration is generally much quicker.
3 *Flexibility* – the parties have more control over the timing and the format of the proceedings. Mutually agreeable arrangements can be made to cause minimum disturbance to normal business.
4 *Technical expertise of the arbitrator* – the Arbitrator will be selected by the parties on the basis of technical expertise and suitability to resolve the dispute in question. Decisions should, therefore, tend to be more in keeping with actual practice in the area under dispute. In litigation, a judge lacking detailed technical knowledge may be appointed.
5 *Privacy* – arbitration hearings are normally held behind closed doors. Sensitive business information is therefore kept private, and the harm that can result from damaging publicity is avoided.

Once the parties to a contract have agreed to settle a dispute by reference to arbitration, this is the method that should be used. Agreement to arbitration can be made at the time of dispute or by an agreement in the contract. In most situations the courts will refuse to hear a case where referral to arbitration has been agreed by the parties involved.

Since the introduction of statutory adjudication in 1998, the use of arbitration as a means of settling construction disputes has diminished, a situation that has been recognized in the current JCT forms of contract. If an Employer wishes to use arbitration as a means of settling a dispute, this fact must be clearly stated in the Contract Particulars under Article 8, failure to do this means that disputes will be referred to litigation, i.e., the default process.

Where under Article 8 it has been stated that disputes are to be referred to arbitration, then the procedures are to be conducted in accordance with the JCT 05 edition of the Construction Industry Model Arbitration Rules (CIMAR).

Clause 9.4.1 requires one of the parties to a dispute to give written notice to the other party to the effect that a dispute or difference will be referred to arbitration. The Arbitrator is to be a person agreed by the parties; if the parties cannot agree within 14 days of the written notice, then a person is to be appointed by the person named in the Appendix to the contract. The notice should identify

(i) the agreement to which the notice relates;
(ii) the matters in dispute and to be referred to arbitration; and
(iii) that the appointment of an agreed Arbitrator is required, but in the absence of agreement the Appointer identified in the contract will be requested to appoint an Arbitrator.

Clause 9.5 gives wide powers to an Arbitrator to determine all matters in dispute. These powers include reviewing and revising certificates, opinions, decisions, requirements or notices. Any award of an Arbitrator is final and binding. However, appeals and applications can be made to the High Court in connection with questions of law. The provisions of the Arbitration Acts apply to arbitrations arising in connection with JCT contracts.

The JCT CIMAR Rules

The JCT/CIMAR Arbitration Rules[1] are intended for use with JCT Standard Forms of Building Contract, and describe the procedure and programme to be adopted in the event of arbitration. Instead of continuing to publish its own set of arbitration rules the JCT has decided to adopt the CIMAR publication, which they have then amended by attaching the 'JCT Supplementary and Advisory Procedures'. CIMAR has been produced to provide an industry standard approach to arbitration and to encourage their use that CIMAR has made provision for contract drafting bodies to add 'advisory or model procedures'. The JCT supplement is made up of two sections. Part A, which outlines mandatory procedures that modify some aspects of CIMAR and must be complied with. Whereas Part B provides information on advisory procedures, which may modify or expand on CIMAR. Part B procedures only apply where the parties, after arbitral proceedings have begun, have expressly agreed to their use.

The party requiring a dispute to be referred to arbitration is termed the *Claimant*, the other party is the *Respondent*. Under the JCT Mandatory Procedures there are three alternative procedures for the conduct of the arbitration, and under Rule 6 the parties are to each provide the arbitrator with a proposal as to which procedure to adopt. Under the JCT timetable, the parties must provide this information within 14 days of the date of notice of the arbitrator's acceptance of the appointment. Within 21 days of acceptance of the appointment the arbitrator will then normally hold a 'procedural meeting' with the parties and direct them as to which procedure is to be followed. A procedural meeting need not be held if the parties and arbitrator agree it is not necessary, and the arbitrator will make his direction based on the written submissions. The alternative procedures available to the arbitrator are

- Rule 7: Short hearing
- Rule 8: Documents only
- Rule 9: Full procedure.

The JCT procedure states that unless the parties have agreed on which of the above Rules is to apply, Rule 8 is to apply by default, unless the arbitrator considers it would be more appropriate to use Rule 9.

The following explanations are based on the assumption that the parties have expressly agreed that the advisory procedures in Part B are to apply.

1. Rule 7 – Short hearing

The procedure requires a hearing to take place within 21 days of Rule 7 becoming applicable. At least seven days before this hearing, any necessary supporting documentation should be identified by the parties and where the documentation is not in the possession of one of the parties, it should be served upon that party. The documents identified should be issued to the Arbitrator at least seven days prior to the hearing. The Arbitrator is to make an award within one month from the date on which the hearing was concluded. The Arbitrator may take longer than one month if he requires, in which case he must notify the parties. Each party to the dispute is required to bear its own costs (except for special reasons at the discretion of the Arbitrator).

The Rule 7 procedure is appropriate where there is no dispute of facts that would require cross-examination and where the dispute arises during the course of a project and a quick, binding award is required to enable the project to proceed.

2. Rule 8 – Documents only

From the date Rule 8 becomes applicable, the Claimant has 21 days to serve a statement of the case. Following submission of the Claimant's statement, the Respondent has 28 days in which to submit a statement of defence, together with any counterclaim. The Claimant has 14 days in which to serve a statement in reply to this defence, together with a defence to any counterclaims where applicable. If the Claimant serves a defence to a counterclaim, the Respondent has 14 days to serve a statement in reply to this defence. Any appropriate documents necessary to support statements should be identified by the Claimant and the Respondent in their statements; any documents fundamental to the claim or defence should be included with the statements. If, at any time in this programme, a party does not serve a statement within the required timescale the Arbitrator notifies the parties that he intends to proceed unless the relevant statement is received within seven days of the notice. The Arbitrator should publish his award within one month from the date that the submission of statements was concluded. As explained in Rule 7 the Arbitrator may take longer than one month if required. If either party fails to serve a statement of the case, the Arbitrator's award will be made on the basis of the material submitted. In some circumstances, the Arbitrator has the power to extend the timescales prescribed under Rule 8. With this type of hearing the CIMAR procedure for costs is to apply and the general principle is that the losing party is to bear the costs. Further advice about costs is provided in Rule 13.

Rule 8 is suitable in situations where the Arbitrator is able to decide on the basis of written statements and there is no dispute over the facts.

3. Rule 9 – Full procedure

The procedure under Rule 9 is similar to that under Rule 8 concerning pleadings and discovery with the exception of time limits. The Claimant has 28 days in which to serve a statement of the case, after which the Respondent has 28 days in which to serve a defence and counterclaim. The Claimant subsequently has 28 days in which to exchange a statement of reply to the defence and 28 days in which to exchange a statement of defence to any counterclaims. If the Claimant serves a defence to a counterclaim, the Respondent has 14 days to serve a reply to the defence. Following this, the Arbitrator will consult with the parties and notify them of the time and place of the hearing. On conclusion of the hearing the Arbitrator should inform the parties of a target date for the delivery of the award, and should endeavour to deliver within that date. Again the CIMAR rules regarding costs are to apply for this procedure.

Rule 9 is appropriate where there is a disagreement between the parties on the facts associated with the dispute and the Arbitrator would require oral evidence, together with cross-examination, to establish the facts.

Role of the Arbitrator

The Arbitrator is empowered to inspect work or materials as necessary including the taking of samples and ordering tests or experiments. In any award, he decides the liability of the parties in respect of the payment of arbitration fees and expenses and on liability for paying the other party's costs (except generally under the Rule 7 procedure). An award must be made in writing, dated and signed by the Arbitrator. The award is to contain sufficient information to demonstrate to the parties how the Arbitrator reached his decision, unless the parties have agreed otherwise or alternatively they have agreed the award. The Arbitrator is required to notify the parties when an award is ready for publication, together with the amount of fees and expenses payable. On payment of the Arbitrator's fees and expenses, the award is normally taken up by the successful party.

Alternative dispute resolution

Alternative dispute resolution is the term used to describe a number of formal, non-adversarial methods of resolving disputes without resorting to the courts or arbitration. The approach was developed in the United States and was introduced to the UK in the 1980s. The processes offer an inexpensive, expeditious and probably more palatable alternative to litigation and arbitration. As with arbitration and litigation, the processes involve a third party. However, unlike arbitration and litigation, the role of the third party is not to sit in judgement, but to act as a neutral facilitator to a negotiated settlement between the parties.

The advantages of alternative dispute resolution are often referred to as the four Cs. These are

1. *Consensus* – this approach requires the agreement of all parties to a resolution of the dispute. The emphasis is on finding a business, rather than a legal or adversarial, solution to the dispute.
2. *Continuity of business relations* – the processes are concerned with resolving disputes within the context of, and without permanently damaging, ongoing business relations.
3. *Control* – resolution of the dispute remains in the control of the parties to it. The parties can concentrate on forging a settlement, which focuses on commercial issues rather than the letter of the law and may thus be less damaging for all parties. Once a dispute is referred to the courts or to arbitration, the parties effectively lose control of the process.
4. *Confidentiality* – the proceedings are not published, and therefore the damage resulting from adverse publicity is avoided.

One of the requirements of alternative dispute resolution is that it must be non-binding. If the process is not working, recourse to litigation or arbitration, as appropriate, must be available. If agreement is not reached, the process will seldom have been a waste of time, effort and expense. It will probably have clarified or narrowed the scope of a dispute and will undoubtedly reduce some of the information gathering and synthesis requirements of arbitration or litigation. Where an agreement is reached using one of the alternative methods of dispute resolution, it should be formalized into a written agreement between the parties. Some of the methods of alternative dispute resolution are described below.

Mediation

Under *mediation*, the parties in dispute select an independent third party to assist them in reaching an acceptable settlement to their dispute. This mediator should be skilled in problem solving and preferably have expertise relative to the dispute in question. The role of the mediator is not to make a judgement of the dispute, but to facilitate a settlement between the parties. The normal process involves the mediator meeting with the parties to agree the format and programme. The sessions usually begin jointly with each of the parties presenting their case, informally, to the mediator. This is followed by private sessions, known as 'caucuses', between the mediator and each of the parties. The mediator's role will be to get each of the parties to focus on their main interests and to get them to move to a common position. The mediator will move between caucuses, often passing offers from one party to the other. As he or she will be in a position of knowledge, confidentiality and impartiality are essential. For the same reason, the mediator is in a position to facilitate, and coax the parties into the best possible negotiated settlement.

The JCT through clause 9.1 raises the possibility that the parties may wish to consider mediation as a means of resolving a dispute. They decided not to incorporate any further procedural information or detail in the contract on the basis that it would be preferable for the parties to make these decisions depending upon their individual circumstances and the nature of the dispute.[2]

Although some contract forms encourage the parties to consider mediation as a means of resolving disputes it is not possible to insist that disputes are initially referred to mediation before they may be referred to adjudication. Such a term is in clear contravention of a party's statutory right to adjudication and would be held invalid (*R G Carter Ltd v Edmund Nuttall Ltd, 2000*).

Conciliation

Conciliation is a similar process to mediation. The distinctions are that a conciliator will actively participate in the discussions between the parties, offering views on the cases put forward. There are no private meetings between the conciliator and individual parties to the dispute. It is more informal than mediation, perhaps aimed at getting the parties to discuss differences. Should the parties fail to reach agreement, it is common for the conciliator to recommend how the dispute should be settled.

Mini-trial

A tribunal, usually comprising senior management from the various parties and chaired by an independent adviser, hears presentations made by all parties to the dispute, who may be represented by lawyers. Witnesses and experts may be called to give evidence. Following the presentations, the senior management enter into negotiations, with the objective of reaching an agreed settlement. The chair's role is to facilitate negotiations, adding suggestions and advice as appropriate to encourage agreement. Should settlement not be reached, the chair may, depending upon the agreed procedure, offer a non-binding opinion. The constitution of the tribunal is important. Members of the tribunal should be senior management who have no direct involvement with the project in dispute and possess the relevant authority for negotiating a business settlement. The chair should have the necessary technical and legal experience particular to the dispute, together with the skills necessary to facilitate a negotiated settlement.

Adjudication

Adjudication is a process aimed at providing a quick solution to a dispute. The process involves an agreed third party giving a decision on disputes relating to a contract as and when they are referred to that third party. The adjudicator, together with the procedures and the types of disputes, which can be referred to adjudication, will all be agreed and stated in the conditions of contract.

The process of adjudication provides a speedy resolution to a dispute, thus allowing the parties to proceed with minimal delay and damage to relationships and the project. It may also reduce the number of disputes taken to arbitration and litigation, as it may give a good indication of what any decision is likely to be using these methods of dispute resolution, and therefore dissuade the parties from taking further action.

Prior to 1998 the use of adjudication in the construction industry was fairly limited. Adjudication procedures had been incorporated into some standard forms of sub-contract, but the procedures were only to be used in disputes relating to set off. However, since 1998 the use of adjudication as a method of dispute resolution has become widespread throughout the industry. This increased use of adjudication is a direct result of the Housing Grants, Construction and Regeneration Act 1996. As a result of the Act it is a statutory requirement that an adjudication provision is incorporated into most construction contracts. The JCT initially complied with this requirement by preparing and incorporating its own adjudication procedure into all its standard forms of contract. In 2005 the JCT took a different approach to adjudication, the detailed adjudication procedures were removed from the contract forms and replaced by a simple statement to the effect that any referral to adjudication is to proceed in accordance with the Scheme (see below).

The Scheme for construction contracts[3]

The Scheme has been prepared and published under the supervision of the Houses of Parliament. The purpose of the Scheme is to provide a set of procedures that comply with the Act and are to be used where a construction contract fails to properly incorporate the requirements of the Act. Therefore, it is not possible for parties to a construction contract to avoid the Act by failing to make proper allowance for it within the contract conditions because in such a situation the Scheme will be applied to the contract. Brief details of the Scheme as it relates to adjudication under JCT are reviewed below.

Part 1 – adjudication

Any party to a contract may refer a dispute to adjudication by giving all other parties to the contract a written notice of adjudication. The notice must provide the following brief information

- details of the dispute and the parties involved,
- the redress or solution that is being sought to the dispute,
- names and addresses of the parties to the contract.

After giving the notice of adjudication, the referring party is to ask the named person to act as adjudicator. The JCT have made provision in the contract particulars to allow the parties to name an adjudicator pre-contract if they wish. If an adjudicator is not named or is unable to act in the dispute the referring party is to ask the nominating body to select an adjudicator. In the contract particulars the JCT provides a list of five potential nominating bodies with the intention that four names should be deleted thereby leaving just one nominating body. If the deletions are not carried out then the referring party may select any one of the nominating bodies. The referring party must give a copy of the 'notice of adjudication' when requesting action from a named person or nominating body. A nominating body is expected to give a referring party the name of a proposed adjudicator within five days of receiving the request. If the nominating body fails to meet this deadline the referring party may agree the appointment of a specific adjudicator with the other party to the dispute or request another nominating body to select an adjudicator. A person requested to be an adjudicator should respond within two days stating whether or not he is able to accept. Again procedures are in place to appoint an alternative adjudicator if the initial process fails.

Once the appointment has been successfully completed the referring party must send the adjudicator a referral notice. The notice must be sent within seven days from the issue of the notice of adjudication. Along with the referral notice the referring party should send relevant documentation, i.e., extracts from contract documents and other documents being relied upon in the dispute. Copies of the notice and documents must also be sent to other parties in the dispute. The adjudicator may ask any party to the contract to provide him with further documents or ask for written statements in relation to the referral notice and its accompanying documents.

The adjudicator is to reach a decision within 28 days from the date of the referral notice. The time period may be extended to 42 days with the consent of the referring party or any time period beyond 28 days, where all the parties to the dispute agree. The adjudicator's decision is binding on the parties until the dispute is settled through litigation, arbitration or by agreement between the parties.

The above comments have only highlighted some of the key aspects of the adjudication procedure set out within the Scheme. For any surveyor who is involved or likely to become involved in adjudication, it is important that they refer to the full details of the Scheme.

Adjudication has been described as rough and ready justice; the adjudicator has a very limited time to familiarize himself with the dispute, project and documentation and as a result mistakes are sometimes made resulting in poor decisions. However, despite the adjudicator's decision being binding on the parties they may, if they wish, still pursue the dispute through litigation, arbitration or agreement to reach a final and binding settlement. Therefore, although the process of adjudication may be imperfect it does allow a swift initial settlement of a dispute, which may defuse a potentially damaging incident affecting working relationships and project performance.

Overview

Since the introduction of the HGCRA in 1998 there has been a steady growth in the use of adjudication as a means of settling construction disputes. Studies undertaken by the Glasgow Caledonian University[4] indicate that during the first year of the HGCRA there appeared to be an initial reluctance to use adjudication. Their survey discovered evidence of adjudication being used on 187 occasions in the first year following the Act. However, as construction firms became more accustomed to the principles of adjudication there was a rapid growth in its use to settle disputes; with the result that in the sixth year of its introduction (2003–2004) there was evidence of its use on over 1500 occasions.

Interestingly the Glasgow Caledonian University Reports indicated that quantity surveyors are a dominant force within the adjudicating bodies and in 2004 accounted for approximately 40% of the number of adjudicators registered with the Adjudication Nomination Boards. The dominance of quantity surveyors acting as adjudicators is perhaps not surprising when approximately 80% of adjudications relate to disputes over payment and value.

Conclusion

Since 1998 there has been a steady growth and acceptance of adjudication as a means of settling construction disputes, although it is generally accepted that adjudication is a somewhat 'rough and ready' method of dispute resolution because of the tight time constraints imposed on the procedures. Despite these criticisms it appears that many parties are accepting the decision of the adjudicators. Where a decision is not being accepted the parties are tending to proceed to litigation to obtain a final and binding decision on the dispute with the result that the use of arbitration as a means of dispute settlement is on the decline.

The use of alternative methods of dispute resolution such as mediation is not widespread in the UK, although the JCT has incorporated it into the standard form as a suggested means of dispute resolution (see clause 9.1). The ICE Minor Works Form makes provision for the use of conciliation procedures.

References

1. JCT/CIMAR, *Construction Industry Model Arbitration Rules* (London: Sweet and Maxwell Ltd, 2005).
2. JCT, *Standard Building Contract Guide* (London: Sweet and Maxwell Ltd, 2005), note 113.
3. The Scheme for Construction Contracts (England and Wales Regulations) 1998, The Stationery Office Ltd.
4. Kennedy and Milligan, *Research Analysis of the Progress of Adjudication, Report No 7* (Adjudication Reporting Centre: Glasgow Caledonian University, August 2005).

Bibliography

1. MARSHALL, E. A., *Gill: The Law of Arbitration* (London: Sweet and Maxwell, 1990).
2. LATHAM, Sir Michael, *Constructing the Team* (London: HMSO, 1994), Chapter 9.
3. MCINTYRE, J., Disputes under review, *Chartered Quantity Surveyor*, November 1991, pp. 16–17.
4. SHIFFER, R. A. and SHAPIRO, E., Introducing the Middlemen, *Contract Journal*, 5 April 1990, pp. 18–19.
5. BINGHAM, A., No losers when commerce triumphs over litigation, *Building*, 14 September 1990, p. 48.
6. DIXON, G., Finding a real alternative, *Chartered Quantity Surveyor*, February 1991, p. 20.
7. ASHWORTH, A., *Contractual Procedures in the Construction Industry, Second Edition* (Essex: Longman Scientific & Technical, 1991), Chapter 28, Arbitration.

Appendix A

Particulars of contract referred to in the examples given in the text

Location of site	28–34, Thames Street, Skinton
Description of works	Five lock-up shops at ground level with ten flats over and basement car park and stores
Contract sum	£1,488,000
Employer	Skinton Development Co., High Path, Skinton
Main contractor	Beecon Ltd., River Road, Skinton
Architect	Draw & Partners, 25 Bridge Street, Skinton
Quantity surveyor	Kewess & Partners, 52 High Street, Urbiston
Basis of contract	JCT Standard Building Contract With Quantities, 2005 Edition

Entries on the contract
Particulars See pp. 393–404

Date of possession 19 September 2005

Completion date 21 May 2007

Contract period 20 months

Abstract of information from Contract Bills

Amounts inserted in Bill No. 1 - Preliminaries	£	£
Management and staff	75,000	
Site accommodation	5,040	
Electric lighting and power	900	
Water	1,377	
Telephones	650	
Safety, health and welfare	7,600	
Cleaning	2,770	
Drying out	2,250	
Security	5,450	
Small plant and tools	3,050	
Earthmoving plant	4,000	
Piling plant	5,000	
Temporary roads	4,350	
Access scaffolding	4,450	
Fencing and hoardings	4,620	
Total		126,507

Amounts inserted in Bill No. 5 – Provisional sums		
Statutory authorities:		
Water main connections	7,500	
Electricity main connection	7,500	
Gas main connection	6,000	
Sewer connection	2,000	
		23,000
Dayworks:		
Labour	1,500	
Addition to net cost of labour 150%	2,250	
Materials	1,000	
Addition to net cost of materials 20%	200	
Plant (as Schedule of Basic Plant Charges)	750	
Addition to net cost of plant 40%	300	
Plant (where schedule not applicable)	200	
Addition to net cost of plant 15%	30	
		6,230
Employer's telephone call charges	750	
Disability access	2,000	
External light fittings	1,558	
Landscaping	2,150	
Contingencies	15,000	
		21,458
Total		50,688

Bill No. 3 – Shops and Flats

Summary

Page		£
26	Groundworks	181,837.93
33	In situ concrete	184,537.67
44	Masonry	155,400.25
69	Structural/carcassing metal/timber	69,930.48
79	Cladding/covering	32,051.34
92	Waterproofing	64,103.00
104	Windows/doors/stairs	125,291.63
109	Surface finishes	52,447.88
112	Furniture/equipment	13,598.22
116	Building fabric sundries	32,051.36
121	Disposal systems	29,137.62
124	Piped supply systems	21,045.45
129	Mechanical heating	84,214.65
134	Communications/security/control	22,325.56
136	Electrical services	32,051.79
140	Transport systems	71,235.00
	Total carried to general summary	**£ 1,171,259.83**

141

General summary

Page		£
15	Bill No. 1 Preliminaries/general conditions	126,507.00
19	Bill No. 2 Demolition	5,531.00
141	Bill No. 3 Shops and flats	1,171,259.83
147	Bill No. 4 External works	40,556.00
151	Bill No. 5 Provisional sums	50,688.00
		1,394,541.83
Add for:		
	Insurance against injury to persons and property as item 5C	10,000.00
	All risks insurance as item 5C	12,920.17
	Water for the Works as item 13A	1,377.00
		1,418,839.00
	Add to adjust for errors	29,161.00
	Total carried to form of tender	1,448,000.00

Appendix B

Articles of Agreement

This Agreement is made the 19 August 20 05

Between **The Employer** A Brown

(Company No. A1934)[1]

of/whose registered office is at Skinton Development Company

High Path, Skinton, Middlesex

And **The Contractor** J E Green

(Company No. 129 6265)[1]

of/whose registered office is at Beecon Ltd

River Road, Skinton, Middlesex

[1] Where the Employer or Contractor is neither a company incorporated under the Companies Acts nor a company registered under the laws of another country, delete the references to Company number and registered office. In the case of a company incorporated outside England and Wales, particulars of its place of incorporation should be inserted immediately before its Company number.
As to execution by foreign companies and matters of jurisdiction, see the Guide.

© The Joint Contracts Tribunal Limited 2005

Recitals

Whereas

First the Employer wishes to have the following work carried out[2]:

Erecting a three-storey block of shops and flats

at 21-34, Thames Street, Skinton, Middlesex

('the Works')

and has had drawings and bills of quantities prepared which show and describe the work to be done;

Second the Contractor has supplied the Employer with a fully priced copy of the bills of quantities, which for identification has been signed or initialled by or on behalf of each Party ('the Contract Bills');

and has provided the Employer with the priced schedule of activities annexed to this Contract ('the Activity Schedule')[3];

Third the drawings are numbered/listed in 462/1 to 462/23

annexed to this Contract ('the Contract Drawings') and have for identification been signed or initialled by or on behalf of each Party[4];

Fourth for the purposes of the Construction Industry Scheme (CIS) under the Income and Corporation Taxes Act 1988, the status of the Employer is, as at the Base Date, that stated in the Contract Particulars[5];

~~**Fifth** the Employer has provided the Contractor with a schedule ('the Information Release Schedule') which states the information the Architect/Contract Administrator will release and the time of that release[6];~~

~~**Sixth** the division of the Works into Sections is shown in the Contract Bills and/or the Contract Drawings or in such other documents as are identified in the Contract Particulars[7];~~

[2] State nature and location of intended works.

[3] Delete these lines if a priced Activity Schedule is not provided.
In the Activity Schedule, each activity should be priced, so that the sum of those prices equals the Contract Sum excluding Provisional Sums, prime cost sums and any Contractor's profit thereon, and the value of work for which Approximate Quantities are included in the Contract Bills.

[4] State the identifying numbers of the Contract Drawings or identify the schedule of drawings or other document listing them, which should be annexed to this Contract, and make the appropriate deletions. The drawings themselves should be signed or initialled by or on behalf of each Party.

[5] The CIS was carried into effect by Statutory Instrument 1998 No 2622 (The Income Tax (Sub-contractors in the Construction Industry) (Amendment) Regulations 1998). Those and other applicable Regulations have subsequently been supplemented and amended.

[6] Delete the Fifth Recital if an Information Release Schedule is not provided.

[7] Delete the Sixth Recital if the Works are not divided into Sections.

© The Joint Contracts Tribunal Limited 2005

The Seventh to Tenth Recitals apply only where there is a Contractor's Designed Portion

Seventh the Works include the design and construction of[8] Lift installations ('the Contractor's Designed Portion');

Eighth the Employer has supplied to the Contractor documents showing and describing or otherwise stating his requirements for the design and construction of the Contractor's Designed Portion ('the Employer's Requirements');

Ninth in response to the Employer's Requirements the Contractor has supplied to the Employer:

- documents showing and describing the Contractor's proposals for the design and construction of the Contractor's Designed Portion ('the Contractor's Proposals'); and
- an analysis of the portion of the Contract Sum relating to the Contractor's Designed Portion ('the CDP Analysis');

Tenth the Employer has examined the Contractor's Proposals and, subject to the Conditions, is satisfied that they appear to meet the Employer's Requirements. The Employer's Requirements, the Contractor's Proposals and the CDP Analysis have each for identification been signed or initialled by or on behalf of each Party and particulars of each are given in the Contract Particulars;

[8] State nature of work in the Contractor's Designed Portion, or delete these four Recitals if not applicable. If the space here is insufficient a separate list should be prepared, signed or initialled by or on behalf of each Party and identified here, either as a specified Annex to this Contract or by its reference number, date or other identifier.

© The Joint Contracts Tribunal Limited 2005

Articles

Now it is hereby agreed as follows

Article 1: Contractor's obligations

The Contractor shall carry out and complete the Works in accordance with the Contract Documents.

Article 2: Contract Sum

The Employer shall pay the Contractor at the times and in the manner specified in the Conditions the VAT-exclusive sum of

One million, four hundred and eighty-eight thousand pounds

(£ 1,488,000.00) ('the Contract Sum')

or such other sum as shall become payable under this Contract.

Article 3: Architect/Contract Administrator

For the purposes of this Contract the Architect/Contract Administrator is

I Draw RIBA

of 25 Bridge Street, Skinton, Middlesex

or, if he ceases to be the Architect/Contract Administrator, such other person as the Employer shall nominate in accordance with clause 3·5 of the Conditions.

Article 4: Quantity Surveyor

For the purposes of this Contract the Quantity Surveyor is

C Kewess FRICS

of 52 High Street, Urbiston, Surrey

or, if he ceases to be the Quantity Surveyor, such other person as the Employer shall nominate in accordance with clause 3·5 of the Conditions.

© The Joint Contracts Tribunal Limited 2005

Article 5: Planning Supervisor

The Planning Supervisor for the purposes of the CDM Regulations is the Architect/Contract Administrator

(or)[9] ____

of ____

____

or, if he ceases to be the Planning Supervisor, such other person as the Employer shall appoint pursuant to regulation 6(5) of those regulations.

Article 6: Principal Contractor

The Principal Contractor for the purposes of the CDM Regulations is the Contractor

(or)[9] ____

of ____

____

or, if he ceases to be the Principal Contractor, such other contractor as the Employer shall appoint pursuant to regulation 6(5) of those regulations.

Article 7: Adjudication

If any dispute or difference arises under this Contract, either Party may refer it to adjudication in accordance with clause 9·2.[10]

Article 8: Arbitration

Where Article 8 applies[11], then, subject to Article 7 and the exceptions set out below, any dispute or difference between the Parties of any kind whatsoever arising out of or in connection with this Contract, whether before, during the progress or after the completion or abandonment of the Works or after the termination of the Contractor's employment, shall be referred to arbitration in accordance with clauses 9·3 to 9·8 and the JCT 2005 edition of the Construction Industry Model Arbitration Rules (CIMAR). The exceptions to this Article 8 are:

- any disputes or differences arising under or in respect of the Construction Industry Scheme or VAT, to the extent that legislation provides another method of resolving such disputes or differences; and
- any disputes or differences in connection with the enforcement of any decision of an Adjudicator.

Article 9: Legal proceedings[11]

Subject to Article 7 and (where it applies) to Article 8, the English courts shall have jurisdiction over any dispute or difference between the Parties which arises out of or in connection with this Contract.

---

[9] Insert name of Planning Supervisor only where the Architect/Contract Administrator is not to fulfil that role, and that of the Principal Contractor only if that is to be a person other than the Contractor. If only regulations 7 and 13 of the CDM Regulations apply, delete Articles 5 and 6 in their entirety.

[10] As to adjudication in cases where the Employer is a residential occupier within the meaning of section 106 of the Housing Grants, Construction and Regeneration Act 1996, see the Guide.

[11] If it is intended, subject to the right of adjudication and exceptions stated in Article 8, that disputes or differences should be determined by arbitration and not by legal proceedings, the Contract Particulars **must** state that Article 8 and clauses 9·3 to 9·8 apply and the words "do not apply" **must** be deleted. If the Parties wish any dispute or difference to be determined by the courts of another jurisdiction the appropriate amendment should be made to Article 9 (see also clause 1·12 and Schedule 5 Parts 1 and 2).

© The Joint Contracts Tribunal Limited 2005

Contract Particulars

Note: An asterisk * indicates text that is to be deleted as appropriate.

Part 1: General

Clause etc.	*Subject*	
Fourth Recital and clause 4·7	Construction Industry Scheme (CIS)	Employer at the Base Date * ~~is a 'contractor~~'/is not a 'contractor' for the purposes of the CIS
Sixth Recital	Description of Sections (if any) *(If not shown or described in the Contract Drawings or Contract Bills, state the reference numbers and dates or other identifiers of documents in which they are shown.)*[12]	
Eighth Recital	Employer's Requirements *(State reference numbers and dates or other identifiers of documents in which these are contained.)*[12]	Lift installation, document ID L1A Annexed to contract
Ninth Recital	Contractor's Proposals *(State reference numbers and dates or other identifiers of documents in which these are contained.)*[12]	Proposals: BL/2005/1A Annexed to this contract
Ninth Recital	CDP Analysis *(State reference numbers and dates or other identifiers of documents in which these are contained.)*[12]	Analysis: BL/2005/1B Annexed to this contract
Article 8	Arbitration *(If neither entry is deleted, Article 8 and clauses 9·3 to 9·8 will not apply. If disputes and differences are to be determined by arbitration and not by legal proceedings, it must be stated that Article 8 and clauses 9·3 to 9·8 apply.)*[13]	Article 8 and clauses 9·3 to 9·8 (*Arbitration*) * ~~apply~~/do not apply

[12] If the relevant document or set of documents takes the form of an Annex to this Contract, it is sufficient to refer to that Annex.
[13] On factors to be taken into account by the Parties in considering whether disputes are to be determined by arbitration or by legal proceedings, see the Guide. See also footnote [11].

© The Joint Contracts Tribunal Limited 2005

1·1	Base Date	July 2005
1·1	Date for Completion of the Works *(where completion by Sections does not apply)*	21 May 2007
	Sections: Dates for Completion of Sections[14]	Section ____: ____________ Section ____: ____________ Section ____: ____________
1·7	Addresses for service of notices etc. by the Parties *(If none is stated, the address in each case, unless and until otherwise agreed and subject to clause 1·7·2, shall be that shown at the commencement of the Agreement.)*[15]	Employer ____________ ____________ ____________ (Fax Number) ____________ Contractor ____________ ____________ ____________ (Fax Number) ____________
1·8	Electronic communications The communications that may be made electronically and the format in which those are to be made[14] *(If none are identified, all communications are to be in writing, unless subsequently agreed otherwise.)*	* ~~are as follows/~~ * are set out in the following document Contract Bills ____________ ____________ ____________
2·4	Date of Possession of the site *(where possession by Sections does not apply)*	19 September 2005
	Sections: Dates of Possession of Sections[14]	Section ____: ____________ 20____ Section ____: ____________ 20____ Section ____: ____________ 20____

[14] Continue on further sheets if necessary, which should be signed or initialled by or on behalf of each Party and then be annexed to this Contract.
[15] As to service of notices etc. outside the United Kingdom, see the Guide.

© The Joint Contracts Tribunal Limited 2005

2·5 and 2·29·3	Deferment of possession of the site (*where possession by Sections does not apply*)	Clause 2·5 * applies/~~does not apply~~ Maximum period of deferment (if less than 6 weeks) is Four weeks
	Sections: deferment of possession of Sections	Clause 2·5 * applies/does not apply Maximum period of deferment (if less than 6 weeks) is[14] Section ____ : ____________ Section ____ : ____________ Section ____ : ____________
2·19·3	Contractor's Designed Portion: limit of Contractor's liability for loss of use etc. (if any)	£ 500,000.00
2·32·2	Liquidated damages *(where completion by Sections does not apply)*	at the rate of £ 500 per day
	Sections: rate of liquidated damages for each Section[14]	Section ____ : £ ________ per ____ Section ____ : £ ________ per ____ Section ____ : £ ________ per ____
2·37	Sections: Section Sums[14]	Section ____ : £ ____________ Section ____ : £ ____________ Section ____ : £ ____________
2·38	Rectification Period *(where completion by Sections does not apply)* *(If no other period is stated, the period is 6 months.)*	Twelve months from the date of practical completion of the Works
	Sections: Rectification Periods[14] *(If no other period is stated, the period is 6 months.)*	Section ____ : ________ months Section ____ : ________ months Section ____ : ________ months from the date of practical completion of each Section

© The Joint Contracts Tribunal Limited 2005

Clause	Item	Entry
4·8	Advance payment *(Not applicable where the Employer is a Local Authority)*	Clause 4·8 * applies/~~does not apply~~ If applicable: the advance payment will be[16] £ 50,000.00 / ______ per cent of the Contract Sum and will be paid to the Contractor on 20 August 2005 ; it will be reimbursed to the Employer in the following amount(s) and at the following time(s) £50,000.00 on 20 February 2006
4·8	Advance Payment Bond *(Not applicable where the Employer is a Local Authority)* *(Where an advance payment is to be made, an advance payment bond is required unless stated that it is not required.)*	An advance payment bond * ~~is~~/is not required
4·9·2	Dates of issue of Interim Certificates *(If none are stated, Interim Certificates are to be issued at intervals not exceeding one month up to the date of practical completion of the Works, or the date within one month thereafter; the first Interim Certificate is to be issued within one month of the Date of Possession.)*	The first date is: 19 October 2005 and thereafter the same date in each month or the nearest Business Day in that month[17]
4·17·4	Listed Items – uniquely identified *(Delete the entry if no bond is required.)*	~~* For uniquely identified Listed Items a bond as referred to in clause 4·17·4 in respect of payment for such items is required for~~ £
4·17·5	Listed Items – not uniquely identified *(Delete the entry if clause 4·17·5 does not apply.)*	~~* For Listed Items that are not uniquely identified a bond as referred to in clause 4·17·5 in respect of payment for such items is required for~~ £

[16] Insert either a monetary amount or a percentage figure, delete the alternative and complete the other required details.
[17] The first date should not be more than one month after the Date of Possession. Where it is intended that Interim Certificates be issued on the last day of each month, the entry may be completed/amended to read "the last day of *(insert month)* and thereafter the last day in each month or the nearest Business Day in that month."

© The Joint Contracts Tribunal Limited 2005

4·19	Contractor's Retention Bond *(Not applicable where the Employer is a Local Authority)* *(Not applicable unless stated to apply and relevant particulars are given below)*	Clause 4·19 * ~~applies/~~does not apply If clause 4·19 applies, the maximum aggregate sum for the purposes of clause 2 of the bond is £ ____________ For the purposes of clause 6·3 of the bond, the expiry date shall be ____________
4·20·1	Retention Percentage *(The percentage is 3 per cent unless a different rate is stated.)*	____________ per cent
4·21 and Schedule 7	Fluctuations Options[18] *(If no Fluctuations Option is selected, Option A applies.)*	Schedule 7: * ~~Fluctuations Option A applies/~~ * Fluctuations Option B applies/ * ~~Fluctuations Option C applies~~
	Percentage addition for Fluctuations Option A, paragraph A·12 or Option B, paragraph B·13	Ten per cent
	Formula Rules for Fluctuations Option C, paragraph C·1·2	Rule 3: Base Month ____________ 20____
	(For Local Authorities only)	Rule 3: Non-Adjustable Element ____________ per cent
	(Unless Part II is stated to apply, Part I applies.)	Rules 10 and 30(i): * Part I/Part II of section 2 of the Formula Rules applies[19]
6·4·1·2	Contractor's insurance – injury to persons or property Insurance cover *(for any one occurrence or series of occurrences arising out of one event)*	£ 10,000,000.00

[18] Delete all but one.
[19] The Part to be deleted depends upon which method of formula adjustment (Part I – Work Category Method or Part II – Work Group Method) is applicable.

© The Joint Contracts Tribunal Limited 2005

6·5·1	Insurance – liability of Employer *(Not required unless it is stated that it may be required and the minimum amount of indemnity is stated)*	Insurance * may be required/~~is not required~~ Minimum amount of indemnity for any one occurrence or series of occurrences arising out of one event £ 1,000,000.00 [20]
6·7 and Schedule 3	Insurance of the Works – Insurance Options[18][21]	Schedule 3: * Insurance Option A applies/ * ~~Insurance Option B applies/~~ * ~~Insurance Option C applies~~
6·7 and Schedule 3 Insurance Option A (paragraphs A·1 and A·3), B (paragraph B·1) or C (paragraph C·2)	Percentage to cover professional fees *(If no other percentage is stated, it shall be 15 per cent.)*	______ per cent
6·7 and Schedule 3 Insurance Option A (paragraph A·3)	Annual renewal date of insurance *(as supplied by the Contractor)*	15 March 2006
6·11	Contractor's Designed Portion (CDP) Professional Indemnity insurance	
	Level of cover *(If an alternative is not selected the amount shall be the aggregate amount for any one period of insurance. A period of insurance for these purposes shall be one year unless otherwise stated.)*	Amount of indemnity required * relates to claims or series of claims arising out of one event/ * ~~is the aggregate amount for any one period of insurance~~
	(If no amount is stated, insurance under clause 6·11 shall not be required.)	and is £ 250,000.00
	Level of cover for pollution/contamination claims *(If none is stated, the required level of cover shall be the full amount of the indemnity cover stated above.)*	£ ______
	Expiry of required period of CDP Professional Indemnity insurance *(If no period is selected, the expiry date shall be 6 years from the date of practical completion of the Works.)*	* 6 years/ * ~~12 years/~~ * ______ years (not exceeding 12 years)

[20] If the indemnity is to be for an aggregate amount and not for any one occurrence or series of occurrences the entry should be amended to make this clear.

[21] Obtaining Terrorism Cover, which is necessary in order to comply with the requirements of Insurance Option A, B or C, will involve an additional premium and may in certain situations be difficult to effect. Where a difficulty arises discussion should take place between the Parties and their insurance advisers. See the Guide.

© The Joint Contracts Tribunal Limited 2005

6·13	Joint Fire Code	The Joint Fire Code * applies/~~does not apply~~[22]
	If the Joint Fire Code applies, state whether the insurer under Schedule 3, Insurance Option A, B or C (paragraph C·2) has specified that the Works are a 'Large Project':	* ~~Yes~~/No[22]
6·16	Joint Fire Code – amendments/revisions *(The cost shall be borne by the Contractor unless otherwise stated.)*	The cost, if any, of compliance with amendment(s) or revision(s) to the Joint Fire Code shall be borne by * ~~the Employer~~/the Contractor
7·2	Assignment/grant by Employer of rights under clause 7·2 *(where Sections do not apply)*	Clause 7·2 * applies/~~does not apply~~
	Sections: rights under clause 7·2 *(Delete the entry if rights are not to apply to each Section or amend if they are to apply to certain Sections only.)*	* Rights under clause 7·2 apply to each Section
8·9·2	Period of suspension *(If none is stated, the period is 2 months.)*	______________________
8·11·1·1 to 8·11·1·5	Period of suspension *(If none is stated, the period is 2 months.)*	______________________
9·2·1	Adjudication[23]	The Adjudicator is ______________
	Nominator of Adjudicator – where no Adjudicator is named or where the named Adjudicator is unwilling or unable to act (whenever that is established)[24] *(Where an Adjudicator is not named and a nominator has not been selected, the nominator shall be one of the nominators listed opposite selected by the Party requiring the reference to adjudication.)*	President or a Vice-President or Chairman or a Vice-Chairman: * ~~Royal Institute of British Architects~~ * The Royal Institution of Chartered Surveyors * ~~Construction Confederation~~ * ~~National Specialist Contractors Council~~ * ~~Chartered Institute of Arbitrators~~

[22] Where Insurance Option A applies these entries are made on information supplied by the Contractor.

[23] The Parties should either name the Adjudicator and select the nominator or, alternatively, select only the nominator. As to the naming of adjudicators, see the Guide.
The Adjudication Agreement (Adj) and the Adjudication Agreement (Named Adjudicator) (Adj/N) have been prepared by JCT for use when appointing an Adjudicator.

[24] Delete all but one of the nominating bodies asterisked.

© The Joint Contracts Tribunal Limited 2005

9·4·1	Arbitration[25] Appointor of Arbitrator (and of any replacement)[26] *(If no appointor is selected, the appointor shall be the President or a Vice-President of the Royal Institute of British Architects.)*	President or a Vice-President: * Royal Institute of British Architects * The Royal Institution of Chartered Surveyors * Chartered Institute of Arbitrators

[25] This only applies where the Contract Particulars state (against the reference to Article 8) that Article 8 and clauses 9·3 to 9·8 *(Arbitration)* apply.
[26] Delete all but one of the bodies asterisked.

© The Joint Contracts Tribunal Limited 2005

Part 2 : Third Party Rights and Collateral Warranties

If such rights or warranties are required from the Contractor and the particulars required by Part 2 (A) to (D) are set out in a separate document, state here the reference number, date and/or other identifier of that document[27] ____________________

(or)

complete the particulars below:

P&T Rights Particulars

(A) Identity of Purchasers/Tenants on whom P&T Rights may be conferred, and whether (in the case of the Contractor) those rights are to be conferred as third party rights (clause 7A) or by Collateral Warranty (clause 7C)[28]

Clauses 7A, 7C and 7E of the Conditions

Name, class or description of person[29]	The part of the Works to be purchased or let	State in each case which of clause 7A or 7C is to apply
Tenant	Shop unit	7A
Purchaser	Flat	7A

(Where no persons are identified by name, class or description either above or in a document identified above, P&T Rights cannot be conferred either under clause 7A (as Third Party Rights) or under clause 7C (by Collateral Warranty – CWa/P&T). If in relation to an identified person it is not stated whether clause 7A (Third Party Rights) or clause 7C (Collateral Warranty) applies, clause 7A shall apply.)

[27] The document should for identification be signed or initialled by or on behalf of each Party and annexed to this Contract.
[28] The Contractor may be required to grant rights either as Third Party Rights or Collateral Warranties. In the case of Sub-Contractors, provision is made only for the grant of Collateral Warranties – see the Guide.
[29] As to the Contracts (Rights of Third Parties) Act 1999 and identification of beneficiaries, see the Guide.

© The Joint Contracts Tribunal Limited 2005

Paragraph of Schedule 5, Part 1 or Clause of CWa/P&T[30]	**(B) P&T Rights from the Contractor**	
1·1·2	Applicability of paragraph/clause 1·1·2	Paragraph/clause 1·1·2[30] * applies/~~does not apply~~
	Maximum liability *(Unless paragraph/clause 1·1·2 is stated to apply and the maximum liability is stated, paragraph/clause 1·1·2 shall not apply.)*	The maximum liability is £ 100,000.00
	Type of maximum liability *(If not stated, it shall be an aggregate limit on liability.)*	* Maximum liability is in respect of each breach/ * ~~Maximum liability is an aggregate limit on liability~~
1·3·1	Net Contribution: Consultants *(If none are specified, these shall be the Architect/Contract Administrator and the Quantity Surveyor (including any replacements), together with any other consultants who agree to give third party rights or collateral warranties (or undertakings in similar terms) to any Purchaser(s) and/or Tenant(s).)*	For the purposes of paragraph/clause 1·3·1[30] 'the Consultants' are: Architect: I Draw; Quantity surveyor: C Kewess; Engineer: J Steele
1·3·2	Net Contribution: Sub-Contractors *(If none are specified, these shall be such as agree to give third party rights or collateral warranties (or undertakings in similar terms) to any Purchaser(s) and/or Tenant(s).)*	For the purposes of paragraph/clause 1·3·2[30] 'the Sub-Contractors' are:

Funder Rights Particulars

Clauses 7B, 7D and 7E of the Conditions	**(C) Identity of Funder in whom Funder Rights may be vested under this Contract** *(If not identified by name, class or description either here or in a document identified at the commencement of Part 2, Funder Rights shall not be required from the Contractor.)*	Commercial Investment plc

[30] The paragraph numbers in Schedule 5 are the same as the clause numbers in the JCT Collateral Warranty.

© The Joint Contracts Tribunal Limited 2005

Paragraph of Schedule 5, Part 2 or Clause of CWa/F[30]	**(D) Funder Rights from the Contractor**	
	Nature of Funder Rights from the Contractor *(If neither clause reference is deleted, clause 7B shall apply.)*	* ~~Clause 7B (Third Party Rights) applies/~~ * Clause 7D (Collateral Warranty) applies
1·1	Net Contribution: Consultants and Sub-Contractors *(Unless otherwise stated, these shall be those specified (or deemed to be specified) under (B) above.)*	______________________ ______________________
6·3	Period required, if other than 7 days	______________________

© The Joint Contracts Tribunal Limited 2005

Collateral Warranties from Sub-Contractors

(E) **If warranties are required from sub-contractors, set out the necessary particulars in a separate document and state here the reference number, date and/or other identifier of that document.[27][31][32][33]**

If or to the extent not included in any such document complete the particulars below:

Clauses 3·7 and 3·9 of the Conditions

Sub-contractors from whom Warranties may be required[31]	State whether clause 7E and/or clause 7F applies[32]	Level of Professional Indemnity insurance required (if applicable)[33]
Piling sub-contractor	7E and 7F apply	

For these purposes and for the purposes of any document identified above, unless otherwise stated:

(i) where clause 7E is stated to apply[32], all Purchasers and Tenants identified at (A) above and any Funder identified at (C) above shall be entitled to Collateral Warranties in accordance with clause 7E[34] from each identified sub-contractor;

(ii) where clause 7F is stated to apply[32], the Employer shall be entitled to a Collateral Warranty as set out in the form specified in the Contract Documents from each identified sub-contractor in accordance with clause 7F;

(iii) **If applicable, the level of Professional Indemnity insurance must be specified**[33]; the basis of that cover shall be whichever applies under the Contract Particulars for clause 6·11;

(iv) if a maximum liability is specified under (B) above, that shall also apply in relation to all sub-contractors' Collateral Warranties unless a lower amount is specified;

(v) "the Consultants" for Collateral Warranties – SCWa/P&T and SCWa/F shall be those stated in (B) above;

(vi) if a period other that 7 days is specified in (D) above, that other period shall apply in clause 6·3 of each Collateral Warranty – SCWa/F.

[31] Employers should be selective in listing the sub-contractors (or categories of sub-contractor) from whom collateral warranties may be required (see the Guide).
[32] It must be stated whether clauses 7E and/or 7F are to apply.
[33] Professional Indemnity insurance applies only where the sub-contractor has design responsibilities. As to cover levels and sub-contractors who maintain Product Guarantee cover only, see the Guide.
[34] The Sub-Contractor Collateral Warranty for a Purchaser or Tenant (SCWa/P&T) and the Sub-Contractor Collateral Warranty for a Funder (SCWa/F), referred to in clause 7E, are documents prepared by JCT.

© The Joint Contracts Tribunal Limited 2005

Execution under hand

As witness

the hands of the Parties
or their duly authorised representatives

Signed by or on behalf of the Employer

A. Brown

In the presence of:

L. Simpson

witness' signature

L Simpson

witness' name

9 Parks Avenue, Skinton, Middlesex

witness' address

Signed by or on behalf of the Contractor

J.E. Green

In the presence of:

L. Simpson

witness' signature

L Simpson

witness' name

9 Parks Avenue, Skinton, Middlesex

witness' address

© The Joint Contracts Tribunal Limited 2005

Execution as a Deed

Executed as a Deed by the Employer

namely[1] ______________________________

(Employer a company or other body corporate)[2]

(a) acting by a Director and the *Company* Secretary/two Directors[4] namely - *(or)*[3]- (b) by affixing hereto its common seal in the presence of

______________________ and

____

(Insert names of signatories)

______________________[5]

Signature Director

______________________[5]

Signature *Company* Secretary/Director

[Common seal of company]

(Employer an individual)[6] ______________________

Employer's signature

In the presence of

Witness' signature ______________________

Witness' name ______________________

Witness' address ______________________

____

Executed as a Deed by the Contractor

namely[1] ______________________________

(Contractor a company or other body corporate)[2]

(a) acting by a Director and the *Company* Secretary/two Directors[4] namely - *(or)*[3]- (b) by affixing hereto its common seal in the presence of

______________________ and

____

(Insert names of signatories)

______________________[5]

Signature Director

______________________[5]

Signature *Company* Secretary/Director

[Common seal of company]

(Contractor an individual)[6] ______________________

Contractor's signature

In the presence of

Witness' signature ______________________

Witness' name ______________________

Witness' address ______________________

____

© The Joint Contracts Tribunal Limited 2005

Index to references to clauses in JCT Standard Building Contract With Quantities

Index

Lightning Source UK Ltd.
Milton Keynes UK
UKOW04f1727280914

239223UK00003BA/21/P